WELCOME TO THE UNIVERSE IN 3D

PRINCETON UNIVERSITY PRESS

PRINCETON & OXFORD

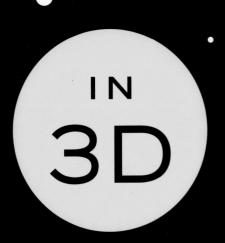

WELCOME
TO THE
UNIVERSE

IN
3D

A VISUAL TOUR

NEIL DeGRASSE TYSON   J. RICHARD GOTT
MICHAEL A. STRAUSS   ROBERT J. VANDERBEI

Requests for permission to reproduce material from this work
should be sent to permissions@press.princeton.edu

Published by Princeton University Press
41 William Street, Princeton, New Jersey 08540
6 Oxford Street, Woodstock, Oxfordshire OX20 1TR

press.princeton.edu

ISBN 978-0-691-19407-3

British Library Cataloging-in-Publication Data is available

Editorial: Ingrid Gnerlich, Arthur Werneck, Whitney Rauenhorst
Production Editorial: Mark Bellis
Text and Cover Design: Chris Ferrante
Production: Steve Sears
Publicity: Sara Henning-Stout, Kate Farquhar-Thomson
Copyeditor: Kathleen Kageff

This book has been composed in Adobe Text Pro
and Trade Gothic LT Std

Printed on acid-free paper. ∞

Printed in China

10 9 8 7 6 5 4 3 2

CONTENTS

PREFACE

Go out and simply look up at the night sky. Everything from the Moon to the planets and stars to the band of the Milky Way appears to be pasted on a two-dimensional surface, the dome of the sky. Yet, over time, humans looking up at the sky (just like you) discovered something fundamental. The universe has depth. By measuring the distances to the objects they observed in the sky, humans began to understand the vast three-dimensional volume of the cosmos. The history of our dawning comprehension of the depths of space is the story of astronomy as a science—the story of humankind's observations and ever-more-accurate measurements of the positions and distances of objects in the universe.

In this book, we will take you on a visual tour of the observable universe—by showing you the universe in depth. We will guide you through a set of spectacular images of the cosmos—of celestial objects and features of the universe that have been observed and measured by astronomers, presented in rough order of their distance from Earth. Each striking image is accompanied by a caption on the facing page, which tells you the story and significance of the image, pointing out interesting features. We begin with the Moon and move outward through planets, stars, and galaxies, finally reaching the cosmic microwave background radiation (CMB), ancient radiation left over from the Big Bang, which is the most distant thing we can see. Light is the fastest thing we know, traveling at 186,000 miles per second. The distances of objects are given in light-travel times—from 1.3 light-seconds for the Moon to 13.8 billion light-years for the CMB. For objects in the solar system, whose distances to us are constantly changing as we and they orbit the Sun, we indicate their distance at their closest point of approach to Earth. These distances, along with highlights of how each object was discovered and measured by astronomers, provide a framework and a narrative thread for the book, which is carried forward from one caption to the next. At each stage of this outward journey, you will learn new and surprising facts about fascinating objects we have found in the depths of space.

You will notice something a bit unusual about the images in this book. Not only will our visual tour of the universe move outward through the three-dimensional depths of space; the images themselves can be viewed in three dimensions, using a special viewer that you will find built into this book. Each pair of images, when viewed with the special stereo viewer (we will describe how to use it a little later on), portrays the celestial object or feature of the universe in three dimensions.

It is our hope that this book will welcome you to the three-dimensional universe, allowing you to see the universe as if for the first time—in true depth. To look at a constellation in 3D and really see that the stars are at different distances reveals to you the volume of space. The familiar coin-like full Moon emerges as a marvelous sphere. Pictures in 3D of the surface of the Moon and Mars make you feel as if you were standing there. We would challenge you to try to understand the orbit of Pluto and other icy Kuiper belt objects *without* looking at a 3D image! And only a stereo view can capture the majesty of the intricate sponge-like pattern of filaments of galaxies connecting galaxy clusters, known as the *cosmic web*. With stereo images, the Hubble Ultra-Deep Field shows a sea of 10,000 galaxies plunging deep into space. Here, on the surface of Earth, we stand on the shore of a vast and deep cosmic ocean. With this 3D book and its stereo viewer, we invite you to dive in and swim in that ocean.

INTRODUCTION

Before we launch our journey through the depths of the cosmos, it is worthwhile to remind ourselves of the fact that it was only through careful observation and measurement of the gleaming and glimmering lights in the sky that human beings realized how distant these lights actually are. Let's explore in a little more detail the methods astronomers came up with to measure distances beyond Earth, stepping out from our own solar system to the furthest galaxies and outward toward the most distant light we can observe, the cosmic microwave background. As we introduce the different methods of measurement, we'll also find ourselves traveling in time, from the third century BC to the current frontiers of astronomy.

How to go about measuring the distances to even the nearest stars is far from obvious. We can't simply stretch a tape measure between here and a star. Humans had to think of another way of measuring distant objects, and we—that is, our insightful ancestors—came up with the concept of *parallax*. Parallax refers to the shifts that occur in the apparent positions of distant objects when your viewing location changes. It is also responsible for your 3D vision. Your two eyes produce slightly different pictures of objects in front of you, due to your eyes' slightly different locations; and these two different images, effortlessly interpreted by your brain, give you depth perception.

You can demonstrate the parallax effect for yourself with a quick experiment. Close your right eye and hold your thumb out at arm's length. Line your thumb up with an object in the distance using your left eye only. Now wink to the other eye. What happens? Your thumb appears to move. Now take your thumb and position it only half an arm's length from your eyes and repeat that exercise. Your thumb shifts even more. The larger the shift, the closer a foreground object is. That's how you judge distance.

Now, let's go back to the problem of how to measure the distance to a star. Obviously, if you use your own eyes to try to measure the distance to a star, you will not be successful. This is because the 2.5 inches between your eyeballs is not enough distance to give you significantly different perspectives on the faraway star. But people realized that they could use the parallax effect for observing stars if they could make observations from two locations sufficiently far apart and then compare them. The distance between one's eyes isn't much—but the diameter of Earth's orbit is 186 million miles. That's a nice wide distance for winking at the universe and deriving a measure of how close a star is.

Let's unpack this a bit further. Think of the nearby star as your thumb, and the diameter of Earth's orbit as the separation between your two eyes. As we know (though this was not immediately obvious to our distant ancestors), Earth orbits around the Sun (see figure 1—not to scale). Earth (the small blue dot) is on one side of the Sun in January and orbits to the other side six months later, in July. In the middle of the figure, there's a nearby star (shown in red), and then way out to the right is a field of much more distant stars.

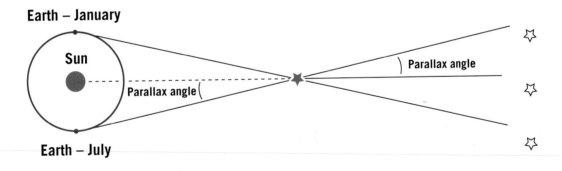

FIGURE 1. Parallax of a nearby star (red) as seen from different locations in Earth's orbit.

Imagine that, in January, we take a picture of the region around the nearby star (see the left of figure 2). We see many, many stars on that photograph, and one of them is the star in question (in red).

Alone, this picture tells us nothing, of course. Remember, we don't know yet which stars are close and which ones are far away. But, say we wait six months and take that picture again in July, from the opposite side of Earth's orbit. Earth has moved to a new position. We now see an identical starry background (as those other stars are much more distant)—but our (red) star appears to have moved from where it once was, to its new location as viewed from Earth in July. It has shifted, just like our thumbs shifted position when we winked our eyes, while everything else in the background basically stayed in the same place (see right side of figure 2).

FIGURE 2. Parallax shifts of a nearby star (red) seen between January and July.

What will happen in another six months? Our red star shifts back to where we saw it a year ago. That shifting just repeats itself, back and forth, over the course of a yearly cycle. Flash the two pictures back and forth, one after the other. If the two photographs are identical except for one star that moves, then we know that *that star* is the one that is closer than all the others. If this star were even closer (like when we moved our thumbs closer to our eyes), then its shift on the picture would be even bigger. Closer stars "shift" more. We put "shift" in quotes because the star is just sitting there—we are the ones moving back and forth around the Sun; the shift is really just due to the change of our perspective when we move from one side of the Sun to the other.

So, now that we understand the concept of parallax, how can it be used to actually measure distance? Note the parallax angle marked in figure 1 next to the nearby star (red). It denotes the angle by which our line of sight from Earth to the nearby star is tipped in the figure relative to horizontal in July. In January our line of sight from Earth to the star is tipped the same amount relative to horizontal in the opposite direction. So during the year, our line of sight oscillates back and forth by plus or minus the parallax angle about a horizontal line. Thus, the total parallax shift seen in figure 2 against the very distant stars (in white) is just two times that parallax angle (shown in figure 1). So if we measure the total shift we see in the sky, we can find the parallax angle. That tells us the shape of the triangle connecting the Sun, the Earth, and the nearby star (red) in July. If that parallax angle is small, the triangle will be very long and skinny, and the nearby star will be much farther away than the Sun. Indeed this turns out to be the case. The parallax angles of even the nearest stars turn out to be less than one second of arc (1/3,600 of a degree). Astronomers thus define a parsec as the distance a star would have if it had a parallax of one second of arc. The word is a mashup of PARallax and SECond of arc—PARSEC. This distance is 206,265 times as large as the Earth-Sun distance. It would take the invention of telescopes to detect parallax for even nearby stars and determine their distances.

But our more distant ancestors had started using parallax long before that, to reveal the true nature of the 3D universe and our place in it. During the total solar eclipse of 190 BC (where the Moon exactly blocked out the Sun as seen in the Hellespont on the Aegean Sea), Hipparchus noted it was recorded that the Moon covered only 4/5 of the Sun's diameter in Alexandria, Egypt. There was a parallax shift of 1/5 the Moon's diameter between the two locations relative to the much more distant Sun. It is like having eyes both in the Hellespont and in Alexandria, whose locations on the globe of the Earth he knew. From that parallax shift, he deduced quite accurately that the distance to the Moon was 60 Earth radii. See the Moon from two rather distant locations on Earth, and you can find its distance. Hipparchus was building on still earlier work.

Aristotle (384–322 BC), the Greek philosopher and polymath, had noted that Earth's shadow on the Moon during a lunar eclipse was always circular in outline. The only shape that always casts a circular shadow from any direction is a sphere. Thus, Aristotle correctly concluded that the Earth was spherical in shape. The Greek astronomer Aristarchus of Samos (310–230 BC), by measuring the size of Earth's shadow as it moved across the Moon in a total lunar eclipse, determined that the Moon was about 1/3 the diameter of Earth.

Aristarchus also made a rough estimate of the Sun's distance from Earth, figuring out that the Sun was larger than Earth and that the Earth orbits the Sun. It just so happens that the Sun and Moon have the same angular size (an object's apparent size in the sky when observed by someone on Earth). That is, they look like they are the same size, if we imagine that they are pasted on a two-dimensional dome of the sky. We knew that their angular diameters are about the same, because we could observe that the Moon just barely covers the Sun during a total solar eclipse. So we also knew that the Sun was farther away than the Moon, but how much farther away?

Aristarchus in addition noticed that, if you look at the Moon at the moment of first quarter, when the Moon is exactly half-illuminated by the Sun as seen from Earth, the rays of the Sun are hitting the Moon from the side, perpendicular to one's line of sight to the Moon. Since the Sun and Moon appear separated by nearly 90° in the sky at this moment, the rays of the Sun hitting the Moon and the Earth at that time must be nearly parallel, and, therefore, the Sun must be much farther away than the Moon. This is illustrated in figure 3.

On the left of figure 3, we see how it would be if the Sun were only twice as far away as the Moon. Since the Moon is exactly half-illuminated by the Sun as seen from Earth, the angle between the Sun and Earth at the Moon must be a right angle (90°). If you were to measure the angle between the Sun and Moon in the sky at this instant, it would be 60° as indicated in the diagram. Euclidean geometry tells us that the sum of the three angles a triangle is 180°. So the parallax angle at the Sun is 30°. (compare with figure 1). In other words, from the angle of 60° between the Sun and Moon in the sky you observed, you could figure out that if you stood on the Moon at that moment, your line of sight to the Sun would be tipped at 30° relative to the vertical in the diagram. You could then make a scale drawing of the triangle and determine that the Sun was two times as far away from Earth as the Moon.

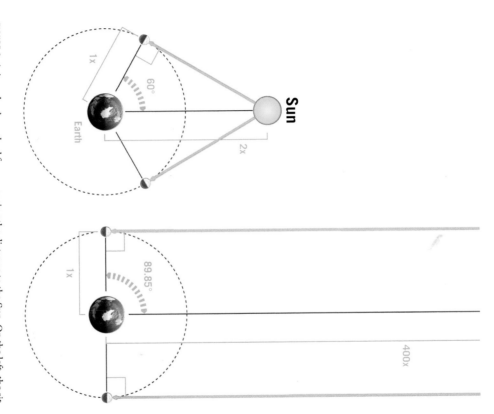

FIGURE 3. Aristarchus's method for measuring the distance to the Sun. On the left, the situation that would occur if the Sun were only twice as far away as the Moon. On the right, the actual situation where the Sun is about 400 times as far away as the Moon.

But that is not what Aristarchus saw. He saw that the Sun was nearly $90°$ away from the Moon in the sky when the Moon was at first quarter. That meant the Sun was very far away, as in the righthand side of figure 3. It was much farther away than the Moon. Compare figures 1 and 3. By his calculations, Aristarchus was effectively putting his eyes at the positions of the Moon at first and last quarter, to get a stereoscopic view of the Sun to determine its distance!

Let's use the actual values. At first quarter, the Moon is $89.85°$ away from the Sun in the sky. This means that the parallax angle at the Sun between the lines of sight from the Earth and the Moon is only $0.15°$. It is a very tall, skinny triangle, and the Sun is far off the top of the page in the right-side diagram. Drawn to scale, the Sun is almost 400 times as far away as the Moon. Now Aristarchus's measurement was not quite as accurate as was required to learn that. He measured the angle between the Sun and Moon at first quarter to be $87°$. Close to the real answer, but not quite. From his measurement he deduced that the parallax angle was $3°$ and that the Sun was 19 times farther away than the Moon. It still would be plotted off the top of the page in figure 3. He was quite correct in asserting that the Sun was much farther away than the Moon, but even then he underestimated its true distance.

Aristarchus then did some remarkable reasoning. Since the Sun and Moon had the same angular size, if the Sun were 19 times as far away as the Moon it must be 19 times the diameter of the Moon. And since he knew the Moon was about 1/3 the diameter of Earth, that meant that the Sun must be 19/3 or over 6 times the diameter of the Earth. The Sun was quite a bit bigger than Earth! Extrapolating from there, he figured that, if the smaller Moon orbits the larger Earth, then the smaller Earth should similarly orbit the even larger Sun. Though his distance estimate was an underestimate, his reasoning was good, and Aristarchus was correct in his conclusion that the Sun is bigger than the Earth (actually it is 109 times Earth's diameter). Through reasoning and measurement, he arrived at the heliocentric model of the solar system—17 centuries before Nicolas Copernicus adopted the same idea!

Sadly, people didn't believe Aristarchus. Unfortunately for Aristarchus (and scientific progress), Aristotle had previously argued that because we do not see the stars showing parallax shifts in a yearly cycle with the naked eye, Earth must be standing still, and does not circle the Sun. He concluded instead that the Earth is at the center of the universe, and that the Sun circles the Earth. Interestingly, Aristarchus had an answer for this. Aristarchus claimed that the stars were infinitely far away, and this was why the parallax was zero. That's not quite right, but he was on the right track! People believed Aristotle, not Aristarchus. It was one of the greatest misses in the history of science. Only in 1543, when Copernicus eventually came along with his book on the subject, were many scientists finally convinced that Earth orbits the Sun.

Why weren't parallax shifts detected in Aristotle's time? Stellar parallax shifts were not observed simply because the shifts are so very small. The naked eye can resolve angular sizes of only around one minute of arc (*arcminute*) or 1/60 of a degree. To get a feeling for how small this is, the Moon has an angular diameter in the sky of about 1/2 degree or 30 arcminutes. As we have noted, the actual parallax angles for even the nearest stars are all smaller than one *arcsecond* (that is, 1/3,600 of a degree)—completely undetectable with the naked eye. In 1543, Copernicus finally responded correctly to Aristotle's argument by saying that the parallax shifts of stars would be undetectable to the naked eye if the stars were simply at immense distances.

It was a long road to vindication for Aristarchus. And it took the invention of telescopes to enable us to actually observe stellar parallaxes. The first star outside our solar system to have its parallax angle measured was 61 Cygni (0.314 seconds of arc) by Fried-rich Bessel in 1838. But this is the way science works. Over generations, over centuries, around the globe, astronomers have continually worked to make better observations and to test and further refine the measurements of those who came before—gradually revealing the reality of our place in the vast cosmos.

Enter Henrietta Leavitt, one of a number of women hired by Harvard Observatory to analyze spectra of stars and classify them, and one of a number of women there who made important discoveries (another was Cecilia Payne who discovered that the Sun was made mostly of hydrogen). In 1912, Henrietta Leavitt discovered that so-called Cepheid variable stars could be used as good "standard candles." A standard candle is a light source whose intrinsic luminosity is known, such as a 100-watt lightbulb. If you see a series of 100-watt lightbulbs in streetlights going down the street, the ones farther away will look dimmer. In fact, we know that the brightness of a lightbulb (or any other standard candle) is lower by a factor of one over the square of its distance, because of the way light spreads as it travels from the bulb. So if you measure the apparent brightness of a 100-watt lightbulb, you can figure out how far away it is. It is thus important to find light sources whose intrinsic luminosity is known: "standard candles."

Cepheid variable stars vary with time in their luminosity, so it might seem odd at first to pick them as standard candles. An individual Cepheid variable star gets more luminous and less luminous in a *regularly repeating pattern* whose time period from peak luminosity to peak luminosity can be measured. Importantly, its overall *average luminosity* (averaged over many cycles of the repeating pattern) can also be accurately defined and measured. It is that average luminosity that makes the star a good "standard candle." By observing a number of Cepheid variables in the Large Magellanic Cloud, a swarm of stars that were all at nearly the same distance from Earth, Leavitt discovered that the longer its period of repetition the larger the average luminosity of that Cepheid variable. Ones with a period of a month had an average luminosity more than ten times greater than ones with a period of three days, for example. Thus, if she could measure the period of a particular Cepheid's variation, she could figure out its average luminosity relative to that of other Cepheids. If you could then measure the distances to a sample of relatively nearby Cepheids (Polaris, the North Star, is one—433 light-years away) using other methods, you could determine the intrinsic luminosities of all Cepheids and, therefore, using Leavitt's work, the distance to any Cepheid variable you saw by simply observing its average brightness and its period of variation. This all took some working out and involved careful identification of the type of Cepheid you wanted, but once it was working you had a set of standard candles. Many Cepheids also happened to be up to more than 10,000 times more luminous than the Sun and so could be seen and identified at great distances. This was a wonderful new way to estimate distances.

In 1922, American astronomer and Rhodes Scholar Edwin Hubble, working at Mount Wilson in California, found Cepheid variable stars in the Andromeda nebula far fainter than any Cepheid variables ever seen before. Using them as "standard candles," he found that this nebula was incredibly distant (the modern value is 2.5 million light-years away). His measurements proved that the Andromeda nebula lies far outside the confines of our Milky Way galaxy and is actually an entire galaxy like our own. In subsequent studies Hubble also discovered that the universe is filled with galaxies, and that the entire thing is expanding. You can think of galaxies in the expanding universe like raisins in a giant loaf of raisin bread, baking in an oven. As the dough expands, the raisins move away from each other. The farther they are apart, the faster they move away from each other. The same thing happens with galaxies. Velocities along our line of sight (which are called *radial velocities*) can be measured by the Doppler effect, the same effect that causes a train whistle to be higher in pitch as it approaches and lower in pitch as it goes away. Spectral lines in stars—particular stellar colors (wavelengths of light) missing from their spectrum because they are being absorbed by chemical elements the stars contain—are Doppler-shifted to the *blue* if the star is approaching us and shifted to the *red* if the star is receding. Hubble found that distant galaxies showed redshifts that were proportional to their distance.

Hubble thus observed that distant galaxies are receding, or moving away from us, and that galaxies twice as far away are moving away from us at twice the speed. This correlation between distance and recessional velocity soon became known as *Hubble's Law*.

By 1931, Hubble and his assistant Milton Humason had extended this relation between distance and recessional velocity to very distant galaxies moving away from us at speeds of up to 45 million miles per hour or 6.7% of the speed of light. This result finally convinced Albert Einstein that the universe is expanding, an idea that he had doubted up to that point. Now, imagine playing a movie of your baking loaf of raisin bread backward in time, and watching the dough contract, so that all the raisins come back toward each other until they all collide. One can imagine a similar time-reversed movie showing the universe contracting from its current state of expansion, so that everything in the universe can be tracked back to a single moment of beginning. This beginning was 13.8 billion years ago, in the form of a "Big Bang." Ever since, the universe has been expanding, with galaxies being flung to ever greater distances from each other. Today, we can use a distant galaxy's redshift (the measurement of how fast it is moving away from us) to estimate its distance to an accuracy of a few percent.

As our observational powers grew (via bigger telescopes, digital cameras, optical and radio telescopes in space), we continued to set our sights on measuring objects farther and farther away, to understand more about the structure and history of the universe. In 1997, two teams—including Adam Riess, Brian Schmidt, and Saul Perlmutter (who were awarded a Nobel Prize for their work in 2011)—found that the expansion of the universe is accelerating. The realization that the universe is expanding ever faster came as quite a surprise. The evidence that proved this came from using *supernovas* (exploding stars that could be calibrated as standard candles just as the Cepheid variables had been before them). But what was causing this startling behavior?

In 1917, Einstein still thought that the universe was static, neither expanding nor contracting. His field equations of general relativity didn't naturally predict a static universe, and to fix this apparent flaw Einstein added a term to his field equations he called the *cosmological constant*. Confronted with Hubble's overwhelming data showing an expanding universe in 1931, Einstein instantly dropped his cosmological constant idea. But in 1934 Belgian Catholic priest Georges Lemaître showed that, in an expanding universe, Einstein's cosmological constant term could still play a role and could be reinterpreted as something we now call *dark energy*. This fills empty space with a *positive energy density* and a *negative* pressure of equal magnitude. The pressure operates in three dimensions (up-down, left-right, front-back) and, being negative, produces a gravitationally repulsive effect. This repulsive effect, according to Einstein's equations, is therefore three times larger than the gravitationally attractive effect inherent to the energy density. This gives dark energy an overall gravitationally repulsive effect. Put another way, the total amount of matter and energy in the universe produces the universe's density—and, for the universe to be expanding at an accelerated rate, something must be counteracting the gravitational attraction of this matter and energy; the gravitationally repulsive negative pressure associated with dark energy nicely does the trick. In fact, by accurately measuring the expansion history of the universe and applying Einstein's original equations of general relativity we can determine the ratio of pressure to energy density in the dark energy; the Planck satellite team finds it to be: -1.008 ± 0.068, which agrees, within the observational errors, with the value of -1 predicted by Lemaître. Remarkable.

The amount of dark energy required to explain the observed acceleration is equivalent (according to Einstein's famous equation $E = mc^2$) to a mass-density of 6×10^{-30} grams per cubic centimeter. This is tiny, not noticeable on small scales—but on cosmic scales, the effects are dramatic. Dark energy is now causing the universe to start doubling in size every 12.2 billion years into the foreseeable future. With such doubling (1, 2, 4, 8, 16, 32, 64, 128, 256, 512, 1,024, etc.), the universe will grow very large indeed. By 122 billion years from now (after 10 doublings), the distances between galaxies should be 1,024 times as large as they are today.

Around 1980, physicist Alan Guth, and others, proposed the idea that the early universe went through an episode of *inflation*, in which an *extraordinarily large* amount of dark energy may have powered the Big Bang itself. In the beginning, there was a rapidly expanding sea of extremely high-energy dark energy, which caused the cosmos to double in size perhaps once every 3×10^{-38} seconds. Bubbles of lower energy would naturally form in this cosmic sea, like bubbles in a pot of boiling water. Each bubble could then turn into an entire universe and expand forever in the continually expanding, inflating sea. In this picture, our universe is just one of many bubble universes. Today we refer to this whole sea of bubble universes as the *multiverse*.

The bubble universe in which we live formed 13.8 billion years ago. Eventually the dark energy inside our expanding bubble decayed into normal particles (in perhaps as little as 10^{-35} seconds) creating a hot Big Bang. Greatly redshifted radiation (light) left over from this hot Big Bang can be seen today as the *cosmic microwave background*. This was discovered in 1965 by Arno Penzias and Robert Wilson working with a radio antenna at Bell Labs tuned to detect microwaves with wavelengths of 2¾ inches (for this they won the 1978 Nobel Prize in Physics). Random quantum fluctuations in such an inflationary bubble universe would be of just the right form to allow growth of the galaxies and clusters of galaxies we observe today. Indeed, these fluctuations are seen in the cosmic microwave background, providing evidence in support of an inflationary beginning of the universe, as Guth and others have proposed. The idea that the early universe began with an epoch of rapidly accelerating expansion (inflation) also gains credence, because, as we have mentioned, we observe a low-energy, slow-motion version of accelerated expansion (inflation) occurring today. And the discovery of the Higgs particle at the Large Hadron Collider in Europe in 2012 has shown us that space must be filled with a Higgs field that gives particles their mass. This suggests that other fields could also fill all the vacuum of empty space, naturally producing dark energy of just the kind proposed by Lemaître in the early 20th century.

As we have seen, astronomers first determined the distances to the Moon, Sun, and planets, then moved on out to measure the distances to the stars. Plumbing even deeper depths of space they found galaxies and the light from the birth of the universe itself. We will now look into these depths to show you the many wonders these astronomers found along the way.

A NOTE ON THE IMAGES

The 3D images in this book are all based on scientific data. They have been produced in an interesting variety of ways. The 3D picture on the surface of the Moon was taken by an astronaut, while the surface of Mars was captured by a two-lens stereo camera on the Mars lander. Other pictures were obtained by the changing viewpoint of a spacecraft camera, as in pictures of topography on the Moon, Venus, and Mars, and in a pair of Hubble Space Telescope pictures of Saturn taken at different times. Pictures in 3D of planets were made by projecting a planet picture onto a sphere mathematically and rocking it by $\pm 3.5°$ to the left and right. For the shot of the surface of Titan, we knew the height of the camera and could calculate the parallax to apply to surface features as they receded toward the horizon. The star charts were produced using parallax data from the Gaia satellite of the European Space Agency (ESA). Pictures of the globular star cluster M13, the galaxy M87 surrounded by its globular clusters, and the Coma cluster of galaxies were produced by applying statistical parallaxes consistent with the two-dimensional locations of objects in the pictures. Jets associated with black holes, along with comet tails, had appropriate parallaxes applied along their lengths. Three-dimensional scientific models and simulations by astronomers were used in some cases, and two appropriate movie frames from simulated flybys were chosen in other cases. Doppler redshift data were used to produce the depth in the pictures of the cosmic web and the Crab nebula.

HOW TO USE THIS BOOK

You will find that the back cover folds out into a second panel with two embedded high-quality lenses. This is called a stereo viewer or a *stereoscope*.

The stereoscope was invented by Charles Wheatstone in 1838. It presents one image to the left eye (showing the left-eye point of view) and a different image to the right eye (from the right-eye point of view). As you'll remember from our earlier discussion of parallax, two images representing slightly different views of an object create the experience of depth—or a 3D visual view. Objects look solid. Sir David Brewster improved and simplified Wheatstone's original design—and we follow Brewster's technique in this book.

If you relax your eyes and look straight forward into the lenses, the two images should fuse into a beautiful 3D image. Most people find this easy to do. If you can't immediately see the 3D effect, be patient and just keep trying, because after a while three-quarters of people having this trouble will suddenly say, "Wow I see it now!" If you are like most people you will be treated to dramatic 3D views.

In the early 1890s, William Friese-Greene patented the first 3D movie system, which projected a left-eye-view movie to your left eye and a right-eye-view movie to your right eye. We salute these pioneers of stereoscopic vision. At WelcomeToThe UniverseIn3D.com, you'll find moving stereo-pair images that you can put on your cell phone and view using the viewer in this book.

Now, we invite you to begin the tour. Hold up the fold-out panel so the lenses face the full color stereo image pairs as you turn the pages, and, once you open to the first image pair, experiment a bit with the position of the lenses. Each person will likely be a little different. Tilting the lens panel slightly will change the focus, to accommodate different eyes, and the lenses will magnify the images to give you a more impressive view. You should see a central fused image in 3D. You may glimpse parts of outrigger flat images to the right and left of the central 3D image. Just ignore these. If you find these distracting, move your eyes backward from the lenses until the side images are blocked from view. You may readjust the viewer and your eyes (resting them occasionally) as you go, viewing objects located at various distances from you in the depths in space. Try it. In a well-lit room, situate yourself to illuminate the images without shadows. Now, relax your eyes, look through the lenses, and a central 3D image should emerge from the stereo pair.

Welcome to the universe—in 3D!

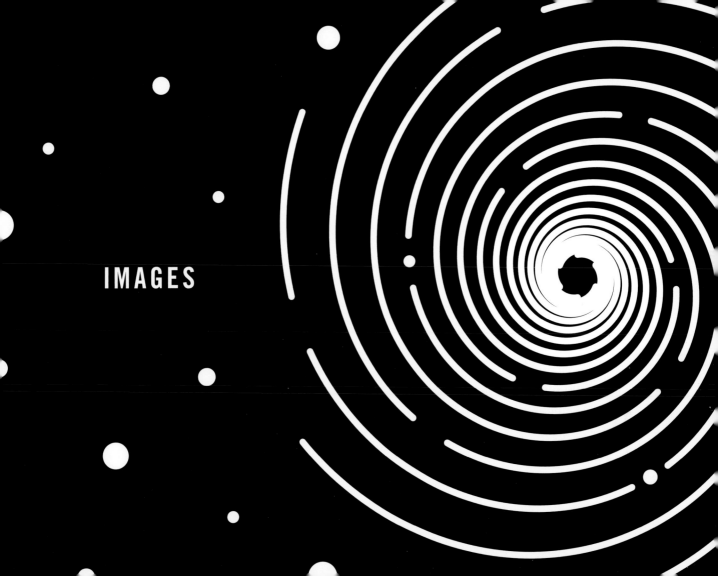

LEAVING EARTH ON A VOYAGE INTO THE UNIVERSE

The surface of Earth marks the shore of a vast cosmic ocean—an ocean we should sail and explore. On February 6, 2018, Elon Musk launched a *Falcon Heavy* rocket from Cape Canaveral in Florida with an unusual payload: a Tesla convertible roadster "driven" by a Starman mannequin wearing a real spacesuit. This stereo image gives you Starman's view of Earth out the front windshield, just as he is about to leave Earth's orbit on his voyage through the solar system out to the asteroid belt. What vistas he will see! The image was constructed by combining two frames from the video of his trip. In this book we will take you even farther into the depths of space than this intrepid Starman will ever go. In the following pages we will travel steadily outward into space, stopping first at the Moon. Objects we encounter along the way will be labeled with their distances from Earth in light-travel time. Welcome aboard!

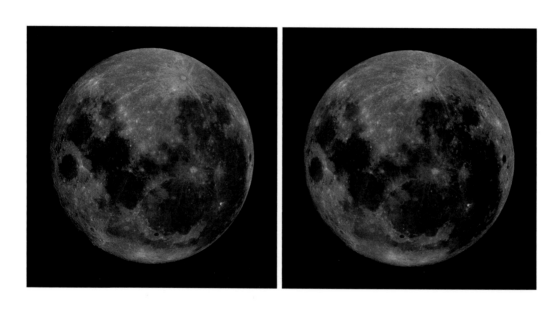

FULL MOON

1.3 LIGHT-SECONDS

When you look at the full Moon, it looks like a flat disk. Why don't you get a sense for its true three dimensions with your naked eyes? Relative to its distance from you—240,000 miles—the distance between your eyes is simply too small to produce a stereo effect. If your eyes were farther apart (say, 30,000 miles apart), you would see the Moon as it appears here: a sphere. Each dark area (*mare*) is an old impact basin where lava flowed long ago and solidified into dark basalt rock. The large, circular *Mare Imbrium* (upper-left quadrant of the picture) was formed by an asteroid impact about 3.85 billion years ago. *Tycho*, the bright crater at bottom-left with debris rays extending from it, was formed about 108 million years ago. The asteroid responsible may have been one of a group, another one of which hit Earth 65 million years ago, causing the extinction of the dinosaurs.

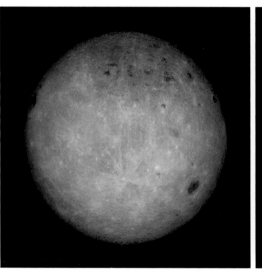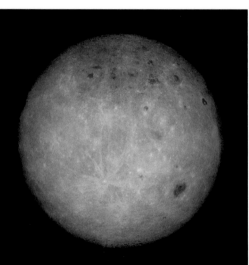

FAR SIDE OF THE MOON

1.3 LIGHT-SECONDS

The Moon is tidally locked with one face always pointed toward Earth. As the Moon orbits, the far side always points away from Earth, and we never see it. The far side was first photographed by the Soviet *Luna 3* spacecraft in 1959, and first seen in person by the *Apollo 8* astronauts when they rounded the Moon in 1968. It has only one small, dark mare, *Mare Moscoviense* (top left). Because the crust on the far side of the Moon is thicker than that on the near side, fewer basalt lava flows occurred when giant impacts occurred, and fewer mares were created. The large patch of gray at the bottom is the South Pole–Aitken basin, the largest impact basin on the Moon (1,600 miles across and 8 miles deep). When we have a new Moon on Earth, the far side of the Moon is fully illuminated by the Sun, as shown here.

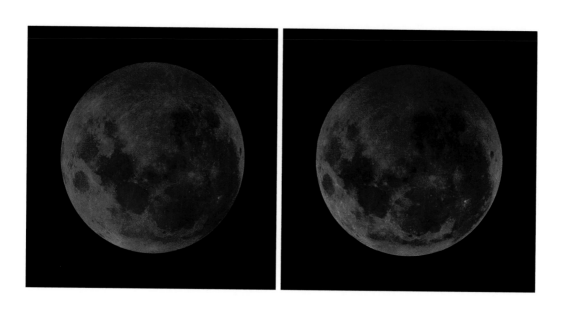

ECLIPSED MOON

1.3 LIGHT-SECONDS

During a total lunar eclipse, the Moon is completely covered by Earth's shadow. Standing on the Moon looking back at Earth, you would see Earth completely blocking out the Sun's shape behind it, but surrounded by a thin, fiery-red ring. As sunlight is bent (refracted) through Earth's atmosphere to reach the lunar surface, it appears red, as in a sunset. This is why the Moon looks red, it is bathed in sunset-colored light that has passed through Earth's atmosphere. The Moon is also *tidally locked*, which means that we only ever see one side of the Moon from Earth. How then could we have taken these two slightly different pictures of the Moon during two different eclipses? As the Moon revolves around Earth in its elliptical orbit, it *librates*, or wobbles slightly in the face it presents to us. This phenomenon allows us to take two images at different times that create a parallax effect and give us a 3D view of the Moon.

MOON

1.3 LIGHT-SECONDS

When lunar astronauts conversed with their fellow humans on Earth, you could notice a 2.6 second delay between a question and the astronauts' answer. The radio waves conveying the conversation travel out to the Moon at the speed of light (1.3 seconds) and then back to Earth (1.3 seconds). This distance of 1.3 light-seconds is the farthest distance human beings have traveled so far. These pictures were taken by *Apollo 17* astronaut Harrison Schmitt in 1972. In the background are the lunar lander (the lunar excursion module or LEM) and the astronauts' lunar roving vehicle or "Moon Buggy."

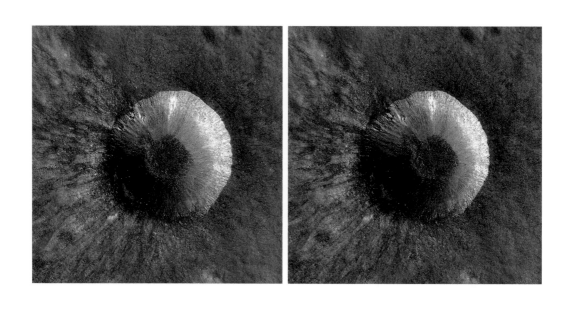

HELL Q CRATER, MOON

1.3 LIGHT-SECONDS

On the Moon's near side, close to Tycho crater (53 miles across, 15,700 feet deep), one can find Hell Q crater. Named after Hungarian astronomer Maximillian Hell (1720–92), Hell Q is much smaller (2.3 miles across and about 3,000 feet deep) and formed more recently than Tycho. The famous 50,000-year-old Barringer crater in Arizona is only 1/3 the diameter and 1/5 the depth of Hell Q crater. On Earth, smaller meteors burn up in the atmosphere before they can impact, and erosion tends to erase old craters left by impactors large enough to have made it to the surface. But on the Moon (which has no atmosphere), all significant impacting objects leave craters that last for eons. These two pictures were taken by the *Lunar Reconnaissance Orbiter* from positions relatively far apart, which exaggerates the vertical 3D effect.

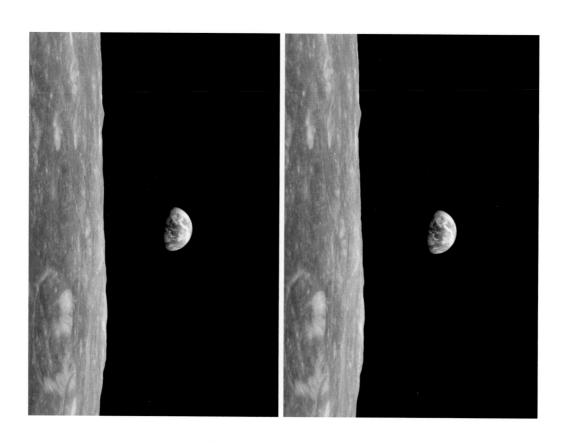

EARTHRISE

1.3 LIGHT-SECONDS

As the *Apollo 8* spacecraft rounded the Moon on December 24, 1968, the beautifully colored Earth emerged from behind the limb (edge) of the Moon and came into view. Astronaut Bill Anders captured "Earthrise" in a photo that has become iconic. Recently, using *Lunar Reconnaissance Orbiter* data, team scientists made a movie to recreate just what those astronauts witnessed as they saw the Earth appear from behind the Moon. We have taken two frames from that movie to make this stereo pair. This allows you to properly appreciate the depth of this awesome view.

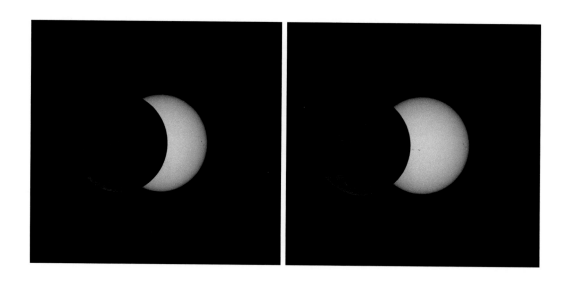

MOON   SUN

1.3 LIGHT-SECONDS   8.3 LIGHT-MINUTES

Here the Moon passes in front of the Sun during the partial phase of the 2017 total solar eclipse. We simulate the near side of the Moon in 3D as if it were illuminated a bit to help you see it, instead of leaving it black. The Moon has moved 1/5 of its diameter between the two pictures, creating a parallax difference and showing the Moon in the foreground in the 3D image. This hearkens back to 190 BC, when the Moon totally covered the Sun as seen from the Hellespont in Turkey but covered only 4/5 of the Sun as seen from Alexandria, Egypt, at the same time. Between the two places, the Moon was shifted by 1/5 its diameter, as in this stereo view. It's like having a stereo view with one eye in the Hellespont and the other in Alexandria! Hipparchus used this difference in views between Alexandria and the Hellespont to make an accurate estimate of the distance to the Moon: 60 Earth radii away.

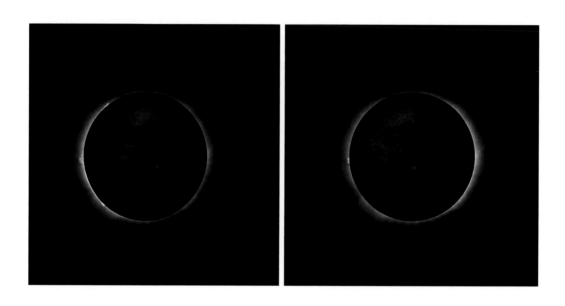

MOON   SUN IN ECLIPSE

1.3 LIGHT-SECONDS   8.3 LIGHT-MINUTES

In this unique stereo view of the total solar eclipse of 2017, the Moon completely covers the bright *photosphere* of the Sun. The Sun's *corona*—its extended, hot, thin upper atmosphere—appears as a white halo. On the right are glowing red *prominences*—plumes of gas ejected along curving magnetic field lines. The left-eye view shows the beginning of totality, while the right-eye view shows the end of totality. In the elapsed time between the two pictures, the Moon has moved about 70 miles, giving a stereo view equivalent to having your eyes 70 miles apart. This produces a slight 3D effect, gently lifting the Moon off the Sun, letting you peek around both sides of it. This reveals the Sun's *chromosphere*—a narrow, hot layer just above its surface. In the picture, this layer looks like a bright, narrow, red line, hugging the upper-right side of the Moon and connecting the prominences.

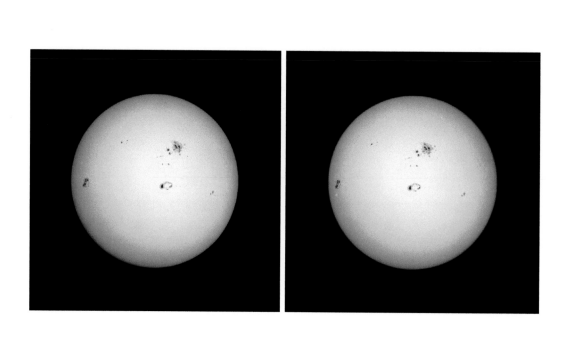

SUN

8.3 LIGHT-MINUTES

Our nearest star is a roiling hot ball of *gas*—109 times the diameter of Earth, 10,000°F at its surface. The rotation of the Sun, faster at the equator than at the poles, twists up the magnetic field lines inside like tightly wound rubber bands. After a while, some of these field lines pop out of the surface as loops, creating pairs of sunspots. You can see a number of sunspots in this colorized picture, created when loops emerged at the surface. The temperature of the sunspots is only 6,400°F. As they are cooler than the rest of the Sun's surface, they glow less brightly, looking dark in comparison to their surroundings.

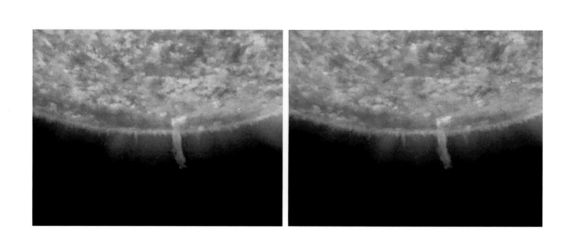

SUN

8.3 LIGHT-MINUTES

This 3D view shows one of the first photographs taken by the twin *STEREO* satellites, launched specifically to create stereo views of the Sun. Taken in March 2007, the picture shows the Sun in the extreme ultraviolet range of the spectrum—that is, at wavelengths of light about a factor of 20 shorter than visible light (the part of the spectrum our eyes can see). Looking at the Sun's short-wavelength ultraviolet light allows us to see the thin, hot regions of the Sun's atmosphere above its visible surface; these regions can reach up to 2.7 million degrees Fahrenheit. The north polar region of the Sun is shown here, and a large *spicule*—jet of gas—is clearly seen extending vertically upward from the Sun's chromosphere into the Sun's corona. The long dark feature at the top is a *coronal hole*, a region associated with open magnetic field lines extending into space along which solar wind particles are ejected.

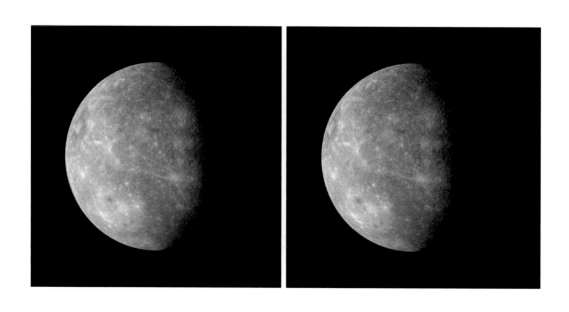

MERCURY

4.4 LIGHT-MINUTES

Mercury, the small rocky planet closest to the Sun, orbits once every 88 Earth-days. It rotates so slowly on its axis that one day on Mercury (from "noon to noon") would take two Mercury-years. Proximity to the Sun produces daylight equatorial temperatures reaching 800°F; however, it is pocked with craters, much like our airless Moon, and water ice has been discovered in the permanently shadowed crater bottoms at Mercury's North Pole. Mercury played a very important role in the history of science, in that careful measurements of its elliptical orbital motion provided an important early test of Einstein's theory of gravity (general relativity).

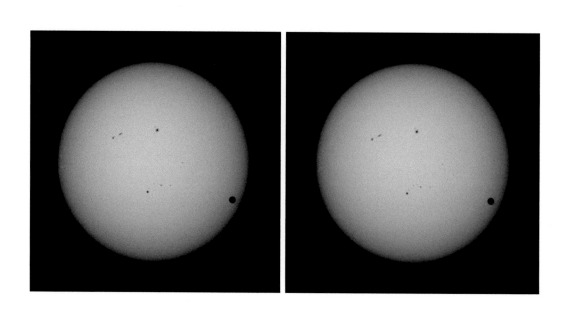

VENUS

2.4 LIGHT-MINUTES

SUN

8.3 LIGHT-MINUTES

Venus, second planet from the Sun, is seen here as a small silhouette in the upper-left quadrant of the Sun. The planet is *transiting* (crossing in front of) the Sun in 2012. The left-eye picture was taken by Bob Vanderbei in New Jersey; the right-eye picture was taken simultaneously by Aram Friedman in Hawaii. New Jersey and Hawaii are a significant fraction of Earth's diameter apart, so the shift of our view of Venus is comparable to the diameter of Venus (since it is nearly the same size as Earth). The observed parallax allowed us to calculate the distance to Venus—26.1 million miles (accurate to 2.8%). We followed in the footsteps of past astronomers. Parallax measurements of Venus made during its 1761 and 1769 transits (including one made in 1769 by Captain Cook in Tahiti) allowed astronomers at the time to estimate the distance to Venus at its point of close approach quite accurately (to within 2.3%).

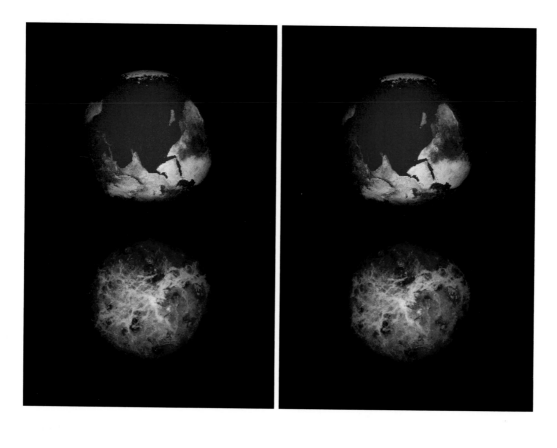

CLOUDLESS VENUS

2.4 LIGHT-MINUTES

CLOUDLESS EARTH FOR COMPARISON

Venus is completely cloud-covered; we have removed the clouds to reveal its surface features (as mapped by radar by the *Magellan* spacecraft, 1990–94). Venus has volcanic plains and two mountainous "continental" highlands. *Aphrodite Terra* (one of the highlands, shown in lighter color) stretches horizontally just south of Venus's equator. Venus's interior heat eventually caused its crust to crack, releasing lava flows that resurfaced the entire planet. The last such event occurred about 500 million years ago. Below Venus, we have shown a cloudless Earth at the same scale. These planets are nearly equal in size but are otherwise quite different. Venus has a thick carbon dioxide (CO_2) atmosphere, with a pressure 92 times that of Earth's mostly nitrogen and oxygen (N_2 and O_2) atmosphere. Venus has sulfuric acid (H_2SO_4) clouds. An extreme greenhouse effect results in surface temperatures of 860°F. In comparison, Earth's abundant oceans, covering 75% of its surface, and its water vapor (H_2O) clouds are much more agreeable!

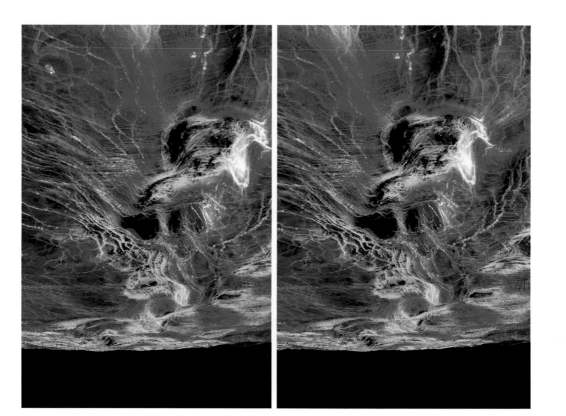

VENUS (SEDNA PLANITIA)

2.4 LIGHT-MINUTES

This image shows the surface altitude features of some Venusian lowland plains, taken by the *Magellan* spacecraft. Areas of higher microwave emissivity from the radar data are shown in orange, and lower-emissivity areas are shown in green and blue. The depressions seen here occur in linear clusters along tectonic belts. A volcanic hot spot causes the raised area, which then cools and subsides to create a central depression. The viewpoint is from an altitude of 200 miles. The terrain is enhanced vertically by a factor of 20 in this picture.

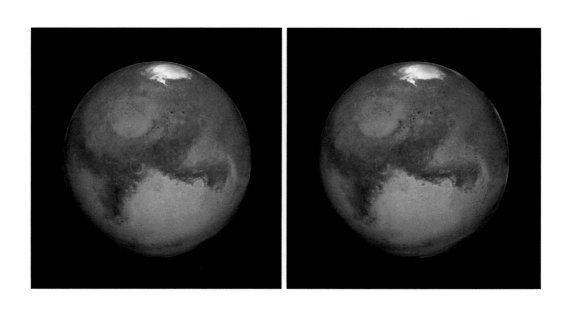

MARS (SYRTIS MAJOR)

3.1 LIGHT-MINUTES

Mars—the Red Planet, the fourth planet from the Sun—gets its color from iron oxide (rust) in the sand and dust covering most of its surface. The most prominent dark area at upper-right center, *Syrtis Major*, is exposed basalt rock not covered by sand and dust. It can be glimpsed in backyard telescopes. Below Syrtis Major, between it and the south polar cap, is the reddish, round desert basin, *Hellas*. It is an ancient asteroid-impact basin, now covered with sand and dust. Hellas contains the lowest-altitude point on Mars: 26,902 feet *below* the mean altitude of the planet. Mars has a carbon dioxide atmosphere with 1/100 the pressure of Earth's atmosphere. Hellas, given its depth, has the most atmosphere above it and so would offer potential Earthling colonists the most protection from the harmful effects of solar flares and galactic cosmic rays. Such intrepid pioneers might live 30 feet below ground and venture out on the surface only occasionally.

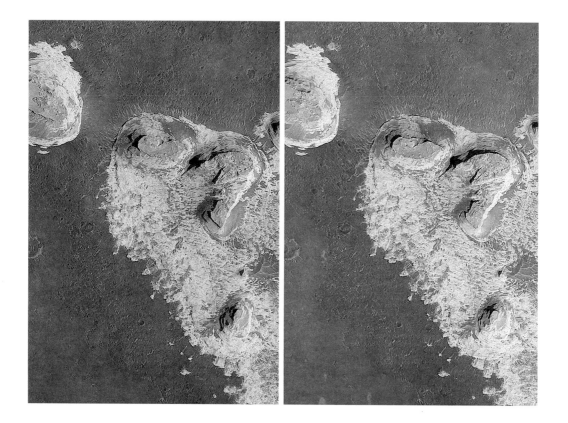

BUTTES ON MARS

3.1 LIGHT-MINUTES

In the northern *Terra Meridiani* area, near the equator of Mars, we can see Martian buttes, which look remarkably similar in appearance to the buttes seen in Monument Valley in Utah. Buttes are produced when a hard rock layer on top shelters softer, more erodible material below. Erosion then occurs, leaving isolated towers with hard rock caps. We believe wind erosion is responsible in this case. The buttes here are about 700 feet tall, and the overall view is about two miles across. This stereo pair combines two photographs taken from orbit by the Mars Global Surveyor satellite on October 23, 2000, and May 15, 2001, to produce an enhanced 3D view.

MARS ROVER

3.1 LIGHT-MINUTES

In 1997, the first Mars rover (*Sojourner*) ventured out on the surface, while a stereo camera on the Mars *Pathfinder* lander (renamed Carl Sagan Memorial Station) captured the historic moment. Having made it successfully down the ramp, *Sojourner* is headed toward the big rock in the picture, nicknamed "Yogi." That's one small roll for a rover, one giant leap for exploring Mars! *Sojourner* was active for 83 *sols* (Mars-days) or 85 Earth-days. In that time it traveled a little over 330 feet, wandering through what appears to be a primeval flood plain, and paved the way for next-generation rovers including *Spirit*, *Opportunity*, *Curiosity*, and *Perseverance*.

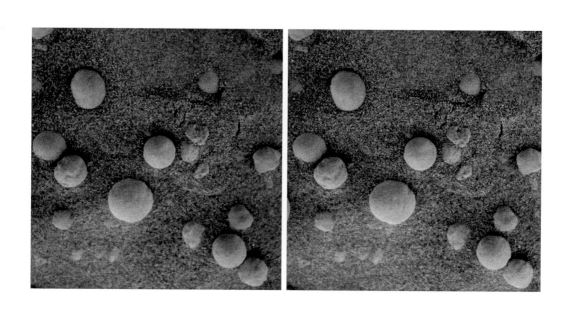

MARS ("BLUEBERRIES")

3.1 LIGHT-MINUTES

Not long after landing on Mars in 2004, the Mars rover *Opportunity* took a picture of tiny, iron-rich hematite (Fe_2O_3) "spherules," measuring 0.4 inches across (shown here in a magnified view). They were nicknamed "blueberries" because they were first shown as blue in false-color NASA pictures. Such mineral inclusions as these can form by accretion in water, and it may be that they present strong evidence that liquid water existed on Mars's surface when they were formed, water that could possibly have supported life. (Note: some crushing of the surface can be seen, where *Opportunity*'s arm touched it to analyze mineral composition.)

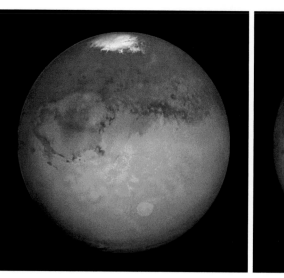

MARS

3.1 LIGHT-MINUTES

Mars's south polar cap (made of water ice with a frosting of carbon dioxide [dry ice]) is easily visible. At the middle right is a roundish "eyeball-like" feature; its central dark spot is called *Solis Lacus*. It was first thought to be a lake, then an oasis connected to a network of canals; we now know it is a plain. The light-colored round spot at the upper-left center is *Olympus Mons*, an extinct volcano with an altitude of 69,841 feet above the mean altitude. Why so big? (Mount Everest, Earth's tallest mountain, is 29,035 feet above sea level.) Olympus Mons is made by a volcanic hot spot, just like some of those on Earth. But on Earth we have moving tectonic plates, so a hot spot on Earth made a string of Hawaiian Islands as the Pacific plate moved. Mars had no crust movement, so the hot spot there made just one enormous volcano. Its base is 370 miles wide.

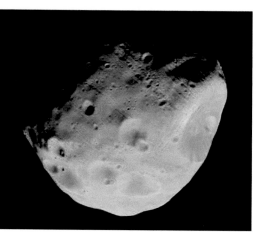

MARS'S MOON PHOBOS

3.1 LIGHT-MINUTES

In his book *Gulliver's Travels* (published in 1726), Jonathan Swift described fictional astronomers' discovery of two tiny Martian moons. Amazingly, in 1877, American astronomer Asaph Hall discovered Mars's two actual moons—Phobos (14 miles in diameter) and Deimos (8 miles in diameter)—and, even more incredibly, their orbital periods were within a factor of two of Swift's fictional satellites! Both highly cratered satellites are thought to be captured asteroids.

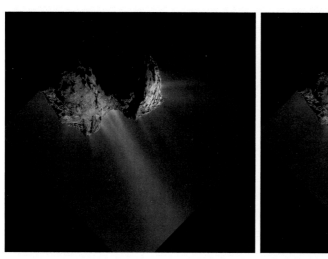

COMET CHURYUMOV-GERASIMENKO

2 LIGHT-MINUTES

Comets are large, dirty snowballs. This comet was trapped in our inner solar system after a close encounter with Jupiter (in 1959), which nudged it into a new orbit, taking it first closer to the Sun than Mars and then sending it out beyond Jupiter. An irregular rubble pile of debris, it apparently formed from the slow collision and sticking of two smaller comets. The lobe on the left is 1.6 miles in diameter, and the one on the right is 2.5 miles across. Its mass is so tiny that gravity has not forced it to coalesce into a round shape. On close approach to the Sun, solar heating of its ice causes outgassing of water vapor and other gases (seen here as multiple jets emitted from its surface) and the release of dust, producing a long "tail."

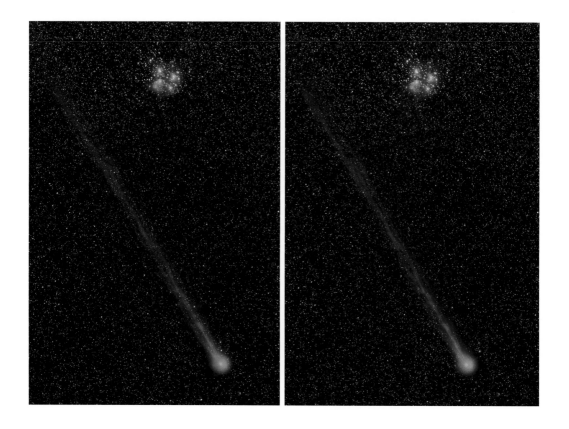

COMET LOVEJOY 2014   PLEIADES

4.9 LIGHT-MINUTES   440 LIGHT-YEARS

In the foreground of this picture is Comet Lovejoy 2014. As it approaches the Sun and heats up, Lovejoy starts to outgas water vapor at a rate of 20 tons per second, accompanied by the release of simple organic molecules such as ethanol (C_2H_6O) and glycolaldehyde ($C_2H_4O_2$). The green glow at its head is caused by fluorescing carbon gas (C_2), while the blue glow from the comet's ion tail is due to carbon monoxide (CO^+), ionized (i.e., stripped of an electron) by ultraviolet solar radiation. Solar wind particles traveling one million miles per hour sweep these ions backward into a dramatic three-million-mile-long tail. This comet, traveling on its highly elliptical orbit, should return to our vicinity in about 8,000 years. In the distant background, you can see the famous *Pleiades* star cluster, made of relatively young, 100-million-year-old, massive blue stars.

ASTEROID VESTA   ASTEROID CERES

9.6 LIGHT-MINUTES   13 LIGHT-MINUTES

The asteroid belt between Mars and Jupiter contains millions of small, rocky asteroids. Shown to scale are the two largest: Vesta (top—diameter 326 miles) and Ceres (bottom—diameter 600 miles). Ceres was discovered first, by Giuseppe Piazzi, in 1801. Vesta, the fourth asteroid to be discovered, looks like a walnut. The bump near its bottom is the central peak of an eight-mile-deep impact crater covering a significant fraction of Vesta's southern hemisphere. Note the more cratered northern hemisphere and the stress furrows (resulting from that big impact) that show up along the equator. This billion-year-old event blasted loose mantle material rich in the mineral olivine, forming a class of meteorites associated with Vesta. Ceres (bottom) is thought to have a rocky core and an icy mantle. Bright areas (colored blue to blue-white for emphasis) are thought to be magnesium sulfate salt deposits from geysers spewing salty water vapor.

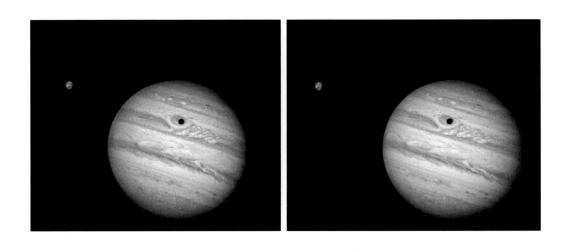

JUPITER AND ITS MOON GANYMEDE

34 LIGHT-MINUTES

Eleven times Earth's diameter, Jupiter is a gas giant composed mostly of hydrogen and helium. Rotating rapidly on its axis once every 10 hours, it is noticeably flattened, like a slightly deflated ball. In the foreground is Ganymede, its largest moon. On April 21, 2014, Ganymede happened to cast its circular shadow on Jupiter's famous *Great Red Spot*, a storm that has raged for hundreds of years. Tinted red by organic molecules, it rotates counterclockwise with winds raging up to 270 miles per hour. Its cloud tops are five miles higher than the surrounding cloud layers. Jupiter has white ammonia (NH_3) clouds, brown ammonium hydrosulfide (NH_4SH) clouds, and water vapor (H_2O) clouds. The prominent brown cloud band north of the equator is called the *North Equatorial Belt*. Lightning on Jupiter can disassociate small amounts of atmospheric methane (CH_4), creating carbon soot that compresses into diamond hail as it falls through Jupiter's deep atmosphere.

MARS AND JUPITER FROM EARTH

3.1 AND 34 LIGHT-MINUTES

Mars and Jupiter are shown here at their points of closest approach to Earth, to illustrate their relative apparent sizes as viewed through a telescope from Earth. Jupiter's cloud belts and Great Red Spot are clearly visible. On Mars, the south polar cap is prominent. Jupiter was photographed from Cebu City in the Philippines, using a 14-inch-diameter reflecting telescope, while Mars was photographed from Chile, using a 39-inch-diameter reflecting telescope. The images were computer-rotated 3.5° to the left and right to create different perspectives for each eye, allowing the stereo pair to produce a 3D view. Mars, being closer, is placed slightly in the foreground. You may cross your eyes a bit to focus on it just as you would for a foreground object in real life.

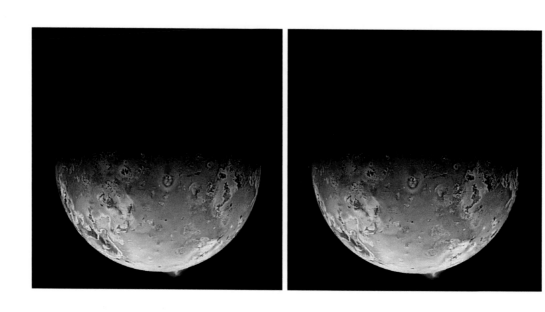

JUPITER'S MOON IO

34 LIGHT-MINUTES

A bit larger than our Moon, Io is the innermost of Jupiter's four large moons. As it circles Jupiter, Io is gravitationally perturbed by its three sister moons, first being pushed slightly closer to Jupiter and then being pulled farther away. As Io gets closer to Jupiter, the tidal force exerted on it by Jupiter strengthens and tends to squeeze it into a slightly elliptical shape. As it moves farther away from Jupiter the tidal force on it becomes weaker, and it relaxes back into a more spherical shape. This constant kneading creates friction, which heats up Io's interior and causes active volcanoes to pockmark its entire surface. An active volcanic plume (blue) can be seen beautifully silhouetted against the black sky above the top horizon. No other body in the solar system today is as violently volcanically active as Io.

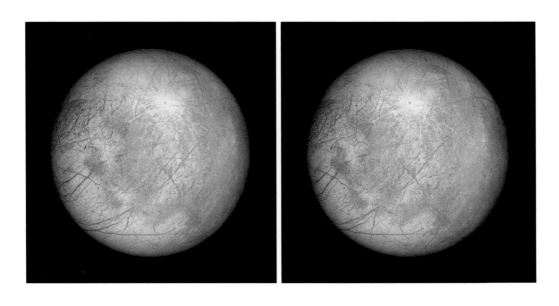

JUPITER'S MOON EUROPA

34 LIGHT-MINUTES

A bit smaller than our Moon, icy Europa is the second innermost of Jupiter's four large moons. Being farther away from Jupiter, Europa has tidal kneading less extreme than that experienced by Io but still considerable. This kneading melts Europa's subsurface ice to form a 50-mile-deep ocean hiding below a 6-mile-deep surface layer of ice. Cracks in the ice layer are visible in this picture, looking much like those in the Antarctic shelf ice and suggesting movement over a mysterious ocean below. There is more water in Europa's ocean than in all the oceans of Earth. One wonders whether life could be lurking there. On Earth, tube worms and other organisms thrive in extreme environments at the ocean floor, on energy supplied by hydrothermal vents called *black smokers*. This could happen on Europa, supporting life-forms isolated from the Sun's energy. A probe could melt its way through the ice layer and release a submarine to explore.

SATURN

1.2 LIGHT-HOURS

Saturn, the sixth planet from the Sun and the solar system's second-largest planet (after Jupiter), is a gas giant nine times the diameter of Earth, surrounded by striking rings composed of icy particles. These particles range from marble size to house size, orbiting the planet in a thin plane less than 2.5 miles thick. One popular theory of how the rings formed is that an errant moon of Saturn wandered close enough to the planet to be torn apart by tidal forces—perhaps less than 100 million years ago. The *Cassini Division*, the 3,000-mile-wide dark gap in the rings easily visible here, was discovered by Giovanni Cassini in 1675. It was formed by one of Saturn's smaller moons, Mimas. Ring particles at the Cassini Division's inner edge circle Saturn twice every time Saturn's moon Mimas orbits once. Mimas's periodic gravitational tugging acts to cause a pile-up of icy particles at that inner edge, and a clearing of icy particles just beyond.

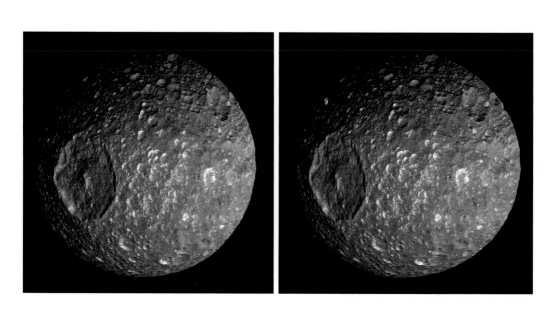

SATURN'S MOON MIMAS

1.2 LIGHT-HOURS

Discovered by William Herschel in 1789, Saturn's icy moon Mimas is 246 miles in diameter and orbits just outside the visible rings. It has an enormous impact crater, discovered in 1980 and named in honor of Herschel, that is 80 miles across, with two-mile-high walls, a floor plunging six miles deep, and a central peak rising four miles from the floor. This crater makes Mimas look rather like the famous Death Star space station in the 1977 movie *Star Wars*. The movie came out three years before the crater was discovered.

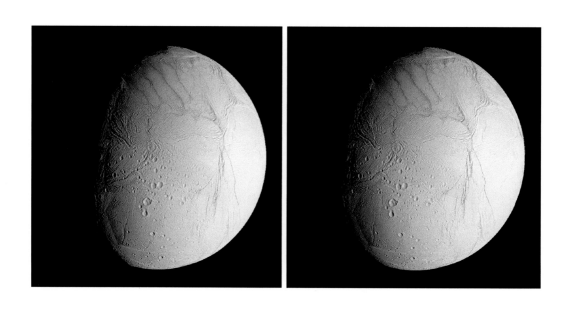

SATURN'S MOON ENCELADUS

1.2 LIGHT-HOURS

Enceladus (approximately 313 miles wide) is much like Jupiter's moon Europa. Tidal kneading from Saturn has melted its subsurface ice to form a southern ocean about six miles deep. As the moon flexes slightly in shape, parallel cracks (shown in blue) appear in the 20-mile-thick layer of sea ice covering this southern ocean. From these cracks, numerous geysers of water vapor outgas into space. This made it possible for the orbiting *Cassini* spacecraft to obtain samples from the subsurface ocean as it flew by. Enceladus's ocean water is salty (NaCl) with traces of complex hydrocarbons including benzene (C_6H_6). Enceladus's ocean (like Europa's) might harbor life. It is interesting that we have had a taste of that ocean already! Some water vapor from the geysers falls back on the moon's surface as snow. The northern hemisphere has many impact craters. However, in the south, the ice is moving over the subsurface ocean and being resurfaced—erasing craters as they form.

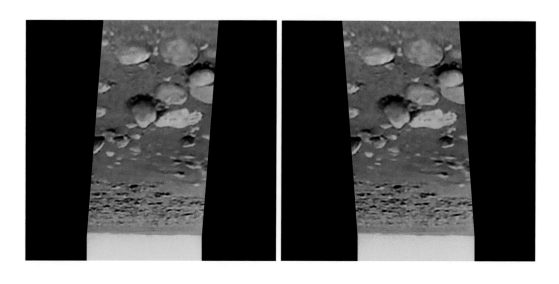

SATURN'S MOON TITAN

1.2 LIGHT-HOURS

Saturn's largest moon, Titan (1.5 times the diameter of Earth's Moon), was discovered in 1655 by Christiaan Huygens. It has a surface temperature of −290°F and an atmosphere with 1.5 times the surface pressure of Earth's atmosphere. While Earth's nitrogen/oxygen atmosphere contains water vapor, water vapor clouds, and water rain, Titan's is composed mainly of nitrogen and contains methane vapor, methane clouds, and methane rain. Titan also has liquid methane lakes and methane rivers that flow out into basins such as the one in this view, then freeze into methane/ethane sand, rich in complex hydrocarbons. The rocks in this photograph (taken by the *Huygens* lander in 2005) are made of water ice. The largest one, lying flat at the lower-left center, is 6 inches across and 33 inches from the camera. Titan is shrouded in thick hydrocarbon smog, giving the scene an orange cast.

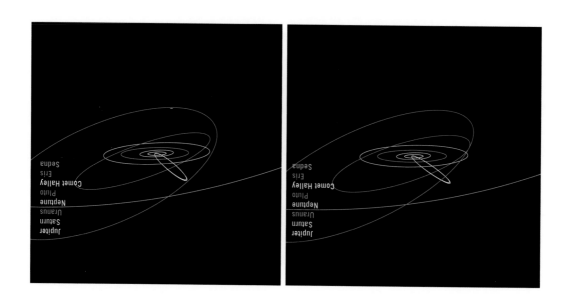

OUTER SOLAR SYSTEM

<10 LIGHT-HOURS

The orbits of the outer planets Jupiter, Saturn, Uranus, and Neptune are nearly circular and contained in the same plane (*coplanar*). In contrast, Halley's Comet (an icy body that outgasses a large tail when it comes close to the Sun every 76 years) is in a highly elliptical orbit (white), tipped 18° relative to the plane of the planets. Icy Pluto's orbit (purple) is also elliptical and tipped 17° relative to those of the planets. In the outer solar system, astronomers have discovered other small icy bodies (*Kuiper belt objects*), such as Sedna (green orbit) and Eris (red orbit), the latter object being more massive than Pluto. Over 2,500 Kuiper belt objects are now known, many with tipped, elliptical orbits. With this new context in mind, in 2006 the International Astronomical Union voted to reclassify Pluto from a genuine planet to a "dwarf planet." Pluto is simply the largest known of a family of similar objects (Kuiper belt objects) in the outer solar system.

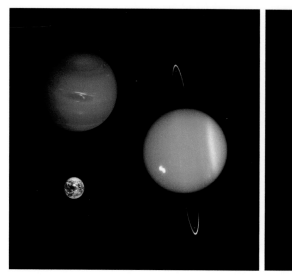

URANUS, NEPTUNE

2.4 AND 4.0 LIGHT-HOURS

WITH EARTH FOR SIZE COMPARISON

Nearly twins in terms of size (and about four times the size of Earth), Uranus and Neptune have other similarities: rocky cores, icy mantles, thick atmospheres of hydrogen and helium, and ammonia and methane clouds. However, Uranus's *rotation axis* is tipped at about 90° relative to the axis of its *orbital plane*, probably owing to some massive impact it suffered in the past. A system of thin rings encircles its equator, shepherded by two small moons. The white spot on Uranus, poking above its cyan cloud layers, may be a methane storm. Neptune's *Great Blue Spot* is a terrific storm; winds of 1,300 miles per hour have been measured on Neptune! Triton, Neptune's largest moon, is a bit smaller than Earth's Moon, but bigger than Pluto. It has nitrogen geysers on its surface shooting five miles high. Triton orbits in a direction opposite to Neptune's spin and may be a captured icy Kuiper belt object.

PLUTO AND ITS MAIN MOON, CHARON

4 LIGHT-HOURS

Pluto (bottom right), slightly smaller than Earth's Moon, is one of many icy *Kuiper belt objects* in the outer solar system. Pluto's frigid surface temperatures range from −402°F to −360°F. The ivory heart-shaped region is named *Tombaugh Regio* after Pluto's discoverer. The heart's left lobe, *Sputnik Planitia*, is composed primarily of nitrogen (N_2) ice. The diagonal dark streak just within the lower left of Sputnik Planitia is an 11,000-foot-high mountain range made of rock-hard H_2O ice, including *Hillary Montes* and *Tenzing Montes*, named after the first climbers to summit Mount Everest on Earth. The dark area cradling the left of the heart at bottom, nicknamed the "Whale," is covered by a complex hydrocarbon tar produced by reactions of Pluto's atmospheric nitrogen and methane with ultraviolet light from the Sun. The dark spot at Charon's northern pole is also tar, produced in a similar way from Pluto's atmospheric gasses, which have drifted over to Charon from Pluto.

OUR SOLAR NEIGHBORHOOD

ALPHA CENTAURI AND SIRIUS

4 AND 9 LIGHT-YEARS

Here is how our solar neighborhood would look from a distance. Our Sun is an ordinary star. Alpha Centauri, appearing here as a single star, is actually a triple star system (a close binary of two solar-type stars orbited at a distance by a smaller red dwarf star, named Proxima Centauri). Proxima Centauri is in turn circled by an Earth-sized planet having an orbit that is 5% the radius of Earth's orbit around the Sun. Barnard's star, 6 light-years from us, is too dim to show up in this picture. Sirius, the brightest star in the night sky, is 9 light-years away and has a white dwarf star companion. Procyon, which also has a white dwarf companion, is 11 light-years away from us. These stars "burn" (fuse) hydrogen in their cores just as the Sun does, except for the white dwarfs, which have exhausted their nuclear fuel and have collapsed down to Earth size.

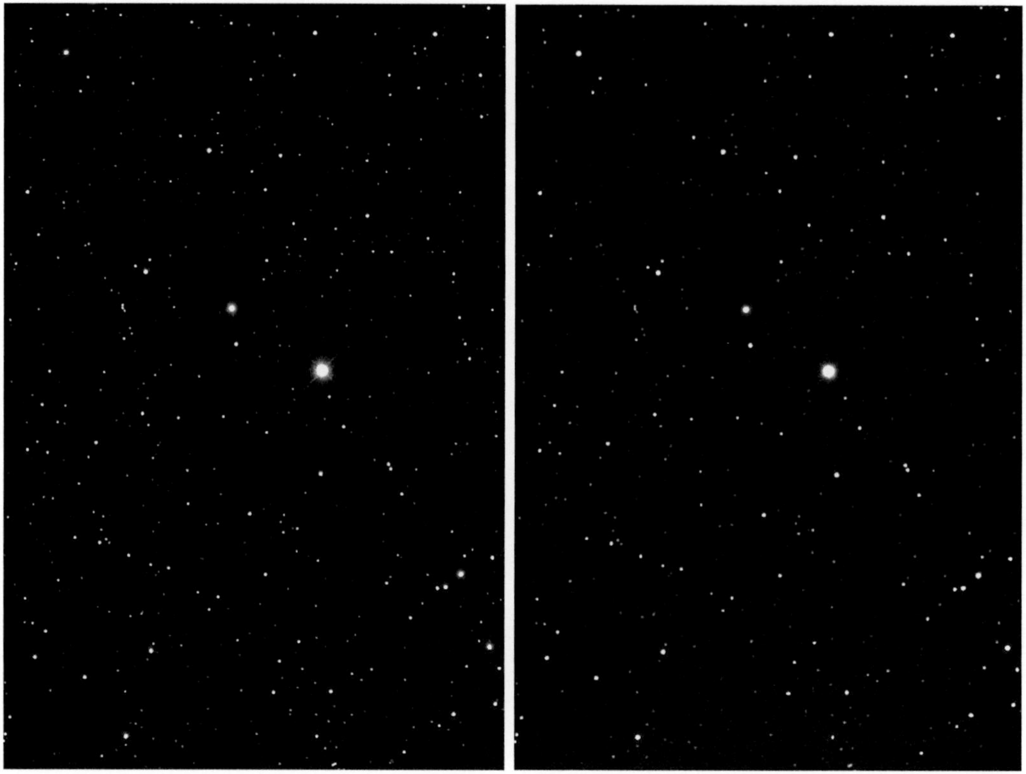

BARNARD'S STAR

6 LIGHT-YEARS

In 1916, E. E. Barnard observed that the star you see in the foreground was moving about 10 arcseconds per year relative to distant background stars. All stars are in motion relative to us, but it usually takes centuries for distant background stars to move noticeably like this. Barnard's star is very close to us and so has a motion that amateur astronomers can detect. In comparison, the background stars here are over 1,000 light-years away. The left picture was taken on June 21, 2012, and the right picture was taken a year later. Earth has returned to the same point in its orbit, so there is no parallax due to Earth's motion around the Sun; however, Barnard's star clearly moves from right to left over time, relative to our Sun. Putting the pictures together gives our eyes a parallax view, and we see Barnard's star as a close object.

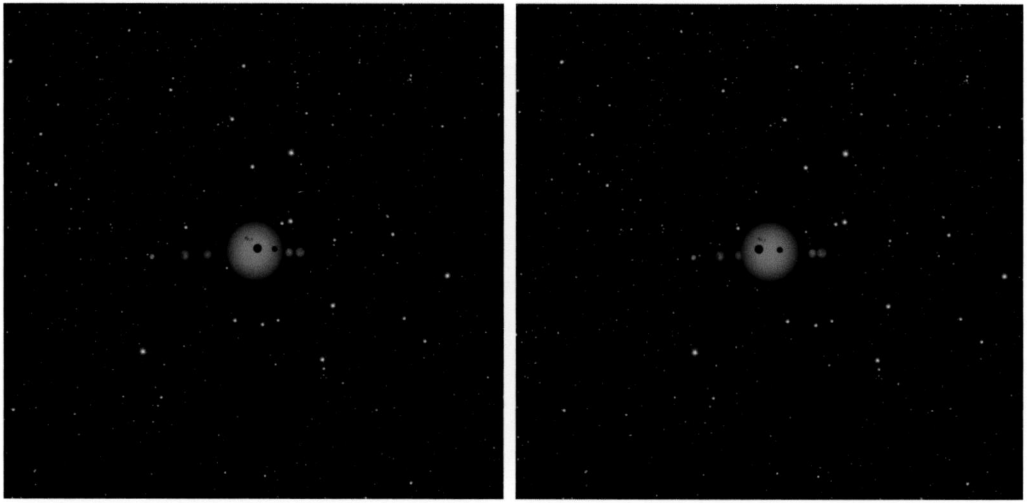

TRAPPIST-1 EXOPLANETS

40 LIGHT-YEARS

We have discovered so far over 4,000 exoplanets orbiting other stars. Trappist-1a is a red dwarf star 12% the Sun's diameter. Cooler (4,000°F) and older (7.6 billion years) than the Sun, it has seven planets (ranging in size from 0.77 to 1.12 Earth diameters), orbiting it in close orbits (between 0.12 and 0.62 Earth-orbit diameters). They are named Trappist-1b, c, d, e, f, g, and h from left to right in this simulated picture. Our line of sight to this system is in the plane of the planets' orbits. These planets were discovered when they transited in front of the star, slightly diminishing its brightness, as d and e are pictured doing now. The other planets, beyond the star, appear red because they are illuminated by a red star. Depending on their atmospheres, several of these planets could have temperatures placing them in the *habitable zone*, where they could have liquid water on their surface. But X-ray and UV radiation from giant flares on Trappist-1a may make them uninhabitable.

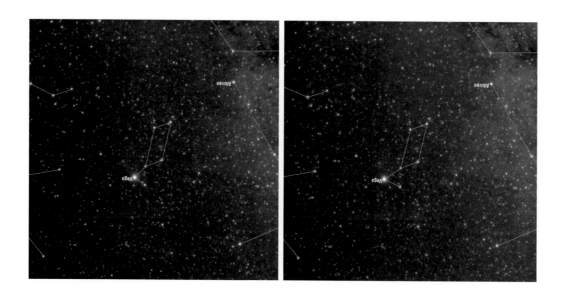

CONSTELLATION OF LYRA

≥25 LIGHT-YEARS

Lyra is a small but prominent summer constellation. Vega—twice as massive as the Sun, twice its diameter, and 40 times more luminous—is its brightest star, and the fifth-brightest star in the night sky. It is also the closest star in the constellation, at a distance of 25 light-years, and so it stands out in the foreground in this stereo view. Between the two photos, the shift of each star is proportional to its parallax shift as seen from Earth during its yearly orbit of the Sun. Vega was the first star to have its photograph taken. The first radio signals from Earth (TV signals from the 1936 Berlin Olympics) reached this star in 1961. If there were any extraterrestrials in its vicinity, they could have sent a return signal arriving at Earth by 1986 (this was used as a plot point by Carl Sagan in his science fiction novel *Contact*).

THE BIG DIPPER

81 TO 124 LIGHT-YEARS

The Big Dipper is the most famous stellar association in the north circumpolar sky, always to be found above the northern horizon, if you are looking at the sky from mid-northern latitudes. The positional differences of stars between the left and right pictures are proportional to their parallax seen from Earth. Constellations are simply two-dimensional patterns we see from Earth. The seven stars forming the Dipper, which are all more massive and more luminous than the Sun, are actually at different distances, as you can clearly see in this 3D view. From left to right, the Dipper stars are Alkaid, Mizar (labeled), Alioth, Megrez, Phecda, Dubhe, and Merak. Most naked-eye stars have Arabic names. After Claudius Ptolemy's second-century *Almagest* (which documented 1,025 stars) was translated into Arabic in the eighth and ninth centuries, Arabic-language names for the stars therein became widely used. Particularly influential was Persian astronomer Abd al-Rahman al-Sufi's richly illustrated book *The Fixed Stars*, published in the year 964.

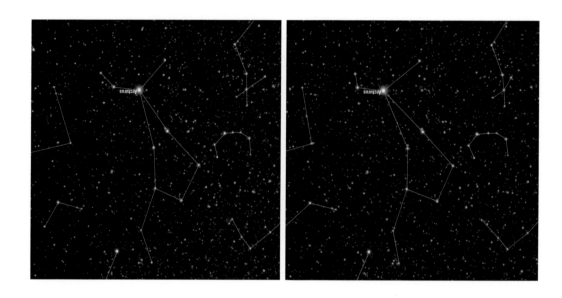

CONSTELLATION OF BOÖTES

≥37 LIGHT-YEARS

The brightest star in Boötes (the Herdsman) is *Arcturus*, a red giant 37 light-years away. Arcturus, 170 times more luminous than the Sun, is also 25 times its diameter. Arcturus is older than the Sun (7.1 billion years versus 4.6 billion years) and 8% more massive. It has run out of hydrogen in its core and has begun puffing up in size, evolving into a red giant. When the Sun does this, as it will 5.5 billion years from now, Earth will burn to a crisp. At the top of this image, you can see the end of the handle of the Big Dipper pointing downward. Find Arcturus in the sky by following the curve of the Dipper handle until you reach it ("arc" to Arcturus!).

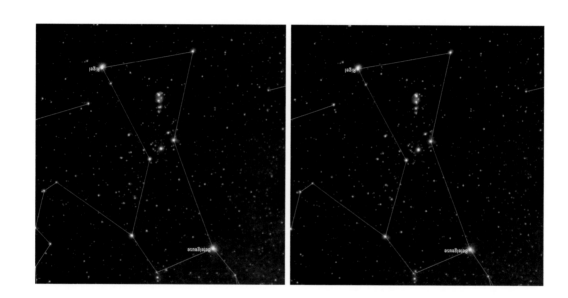

CONSTELLATION OF ORION

BETELGEUSE
640 LIGHT-YEARS

RIGEL
860 LIGHT-YEARS

Orion (the Hunter) is the most spectacular of constellations viewable throughout the world—in the winter from the Northern Hemisphere and in the summer from the Southern Hemisphere. Its two brightest stars appear near the back of this stereo view. The three center stars forming his belt are part of an association of massive *blue* stars. Hanging straight down from the central star of his belt is Orion's sword, where you will find a blurry smudge. This is the Orion nebula—a stellar nursery, where infant stars are being born from gravitationally collapsing clouds of gas and dust. One shoulder is the *red supergiant* Betelgeuse: its diameter is 900 times that of the Sun, and it is about 15 times as massive. Orion's knee, Rigel, is a *blue supergiant*, about 80 times the Sun's diameter and 20 times the Sun's mass. Betelgeuse and Rigel are old stars in their death throes, having exhausted most of their useful nuclear fuel.

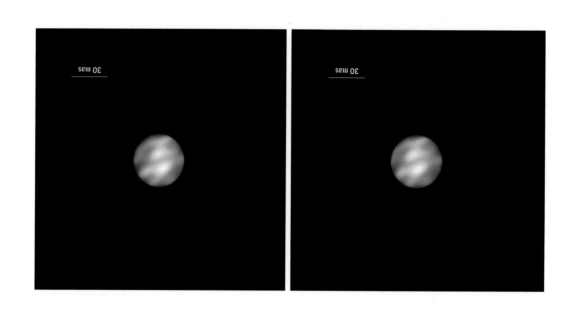

BETELGEUSE

640 LIGHT-YEARS

Betelgeuse is so large that, even though it is far away, we can capture an image of it by jointly utilizing three widely separated telescopes to achieve high resolution. The bar indicates an angular size of 30 milli-arcseconds or 1/120,000 of a degree. The large-scale bright regions are where hotter gas is rising to the surface. Betelgeuse was born as a blue star, burning hydrogen in its core. After exhausting that, it started burning helium into carbon and oxygen in its core, and burning hydrogen in a shell. It will likely end its life as a supernova, leaving behind a 1.4 solar-mass neutron star remnant. Einstein's relativity theory says you can time travel to the future: if you flew to Betelgeuse and back at 99.995% the speed of light, you would age 12.8 years, but you would find Earth 1,280 years older when you returned.

ORION NEBULA

1,400 LIGHT-YEARS

The Orion nebula is a stellar nursery where stars are being born now. The tiny bright group of young stars at the lower center is called the *Trapezium*. Its 40-solar-mass, brightest star is only 2.5 million years old. The upper, clear, star-filled region contains ionized hydrogen (H^+), stripped of its electron by ultraviolet light radiating from the young stars, while neutral hydrogen gas below glows like a fluorescent lamp. Atoms of hydrogen, excited by the ultraviolet light from the young stars, glow with red hydrogen-alpha light as energized electrons drop back to a lower-energy state. New stars are forming downward from the top, eating away the neutral gas at the bottom. Pictures taken by the Hubble Space Telescope have allowed astronomers to create 3D models of this region. This stereo pair is made from two frames of a movie simulating a "space-ship" flyby into the heart of the nebula made by the Hubble Heritage Team.

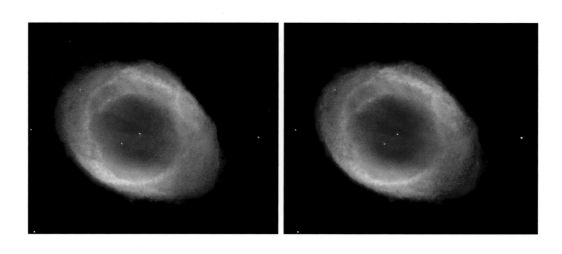

RING NEBULA

2,000 LIGHT-YEARS

The Ring nebula can be seen with a small telescope in the constellation Lyra. After exhausting the hydrogen in its core, a solar-type star starts burning hydrogen in a shell and bloats out to become a red giant—a hundredfold larger than before. Then it expels its outer envelope, as in this picture, revealing its dense core in the center—now a white dwarf star. White dwarfs (formed from stars below a certain mass) are held up against further gravitational contraction because electrons resist being squeezed together, owing to quantum effects. Here, we are looking down on the star's pole, watching an equatorial ring of gas being ejected. The white dwarf emits ultraviolet radiation causing the gasses in the expanding envelope to fluoresce. The blue-green color is from oxygen, and red is from hydrogen. This nebula gives us a preview of what will happen to the Sun when it dies, 7.5 billion years from now.

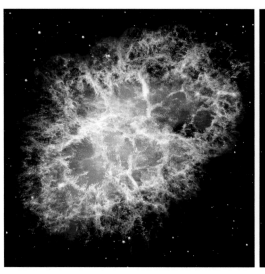

CRAB NEBULA

6,500 LIGHT-YEARS

The Crab nebula is debris from a supernova explosion (death of a massive star) observed by Chinese astronomers in AD 1054, a "guest star" bright enough to be seen in daytime. From pictures taken decades apart, we can measure the apparent motions of the filaments (moving outward from the center). If we trace this motion backward in time, we arrive at approximately AD 1054. At the center now is an 18-mile-diameter neutron star, rotating 30 times per second on its axis. Light beams emitted from its north and south magnetic poles sweep by Earth like a rotating lighthouse beacon. This neutron star is supported against gravitational collapse because neutrons resist being squeezed together owing to quantum effects. Heavy elements from such explosions are dumped back into interstellar gas, from which new stars can form. Gold and other elements heavier than iron can arise from neutron star/neutron star collisions.

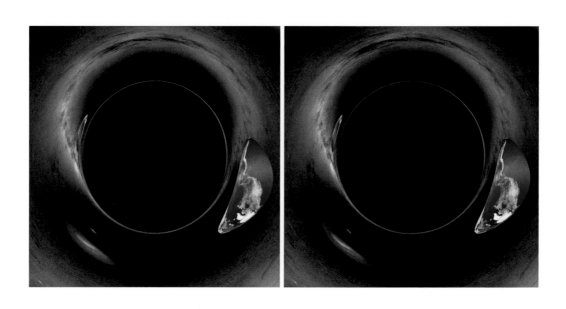

BLACK HOLE

This computer simulation illustrates gravitational light-bending effects around a non-rotating black hole (the black disk)—the collapsed remnant of a massive object. Inside the black hole, gravity is so strong it crushes matter to a point at its center, allowing nothing, not even light, to escape. Earth is located behind the black hole, yet you can see two distorted images of Earth, as light beams bend around both sides of the hole as they travel toward your eyes. Curiously, with 3D vision, these images seem to hover a bit closer to you than the hole itself. The 4,250-solar-mass hole depicted here would have tidal forces that would quickly shred and crush Earth, were it actually to venture this close to the hole. Some black holes, with masses of 4–40 solar masses (radii of 7–70 miles) are formed from collapsed cores of dead massive stars. Others, in the centers of galaxies, have masses of half a million to 10 billion solar masses or more (radii of 1 million to 20 billion miles or more).

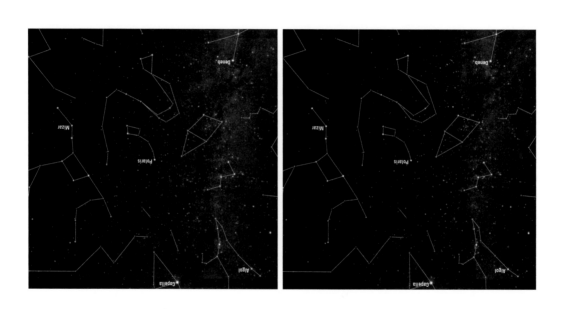

NORTH CIRCUMPOLAR STARS

40 TO 40,000 LIGHT-YEARS

These stars always appear over the northern horizon from mid-northern latitudes, circling the North Star (Polaris) counterclockwise. The Big Dipper appears at the right, facing off against W-shaped Cassiopeia at the left. If you follow the two pointer stars forming the end of the Dipper's cup and keep going past Polaris, you will arrive at Cassiopeia. In the middle, the Little Dipper hangs off Polaris. Deneb, a blue supergiant star near the back, 2,600 light-years away, marks the top of the Northern Cross. Capella in the foreground, by contrast, is relatively close at 40 light-years distance. The band of the Milky Way can be seen at the back left. These are distant stars in the disk of our galaxy, up to 40,000 light-years away. The parallax of a star in this and the following five star charts is proportional to the parallax observed from Earth, giving a view as if your eyes were separated by about 1/5 of a light-year.

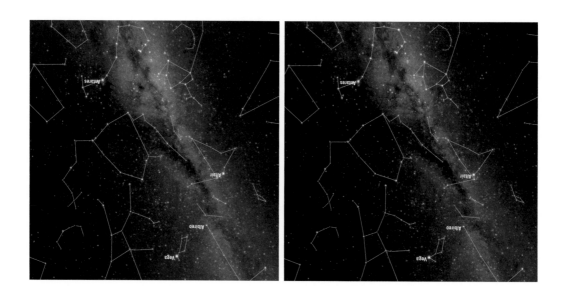

SUMMER STARS

25 TO 75,000 LIGHT-YEARS

From mid-northern latitudes, these stars can be seen at midnight on June 21, hanging over the southern horizon. The stars Vega (in the constellation Lyra) and Altair (in the constellation Aquila) are prominent here. The constellations Sagittarius (which looks like a teapot) and hook-shaped Scorpius (the Scorpion), which contains the red supergiant star Antares, appear at the bottom. Up to 75,000 light-years away, the Milky Way appears as a faint, distant band of light, obscured in part by dark interstellar dust clouds. The Milky Way widens in the region of the galactic center in the constellation of Sagittarius. At the exact center (26,000 light-years away) lurks a 4-million-solar-mass black hole. As seen from a dark site, this is a particularly beautiful region of the sky to observe.

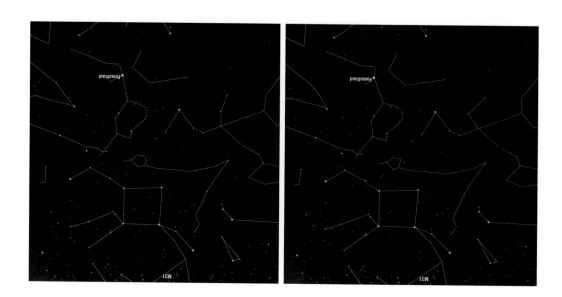

AUTUMN STARS

25 TO 2,500,000 LIGHT-YEARS

From mid-northern latitudes, these stars can be seen at midnight on September 21, hovering over the southern horizon. We are looking out of the plane of our galaxy in this view and see only local stars. Fomalhaut is 25 light-years away. The great square of Pegasus (the Flying Horse) can be seen at center top. He is flying upside down with his head extending to the lower right. The constellation Andromeda is V-shaped off the square toward the upper left. Andromeda contains the Andromeda galaxy (M31), visible here as a fuzzy tilted ellipse near the top center. A companion galaxy to our own Milky Way, it appears small and faint only because it is 2.5 million light-years away. It is in fact about the same size as our Milky Way galaxy.

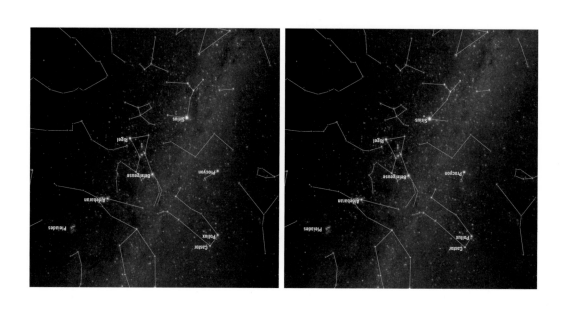

WINTER STARS

9 TO 25,000 LIGHT-YEARS

As seen from mid-northern latitudes, these stars are visible at midnight on December 21, hovering above the southern horizon. Orion dominates the center, with the bright stars Betelgeuse and Rigel at his shoulder and knee. Canis Major contains Sirius (the Dog Star); it is in the extreme foreground, only nine light-years away. Taurus (the Bull) is to the upper right of Orion, with its red eye (Aldebaran). The Pleiades (or Seven Sisters) is the tiny cluster of stars to the right of Taurus. Gemini (the Twins), containing Pollux and Castor, is the constellation above and to the left of Orion. The Milky Way is the diagonal band at the back, whose light comes from myriad stars in our galaxy, ranging up to 25,000 light-years away.

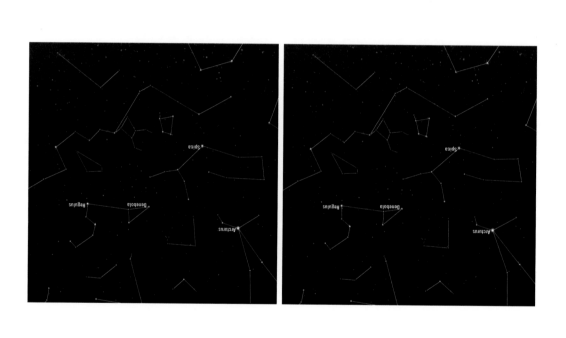

SPRING STARS

37 TO 250 LIGHT-YEARS

As seen from mid-northern latitudes, these stars can be seen at midnight on March 21, hovering over the southern horizon. Leo (the Lion) appears here with his haunches made by a triangle of stars (including Denebola), and his shoulders and head made by a backward question mark of stars starting at the star Regulus. Regulus (77 light-years away) lies within half a degree of the *ecliptic*, the path the Sun traverses in the sky during the year. Venus passed in front of (*occulting*, or blotting out) Regulus on July 7, 1959, and will do it again on October 1, 2044. Virgo (the Virgin), containing the star Spica (250 light-years away), reclines below and to the left of Leo. The lower part of Boötes, containing the bright red giant Arcturus, appears at upper left. We are looking out of the plane of the Milky Way and therefore seeing only relatively nearby stars.

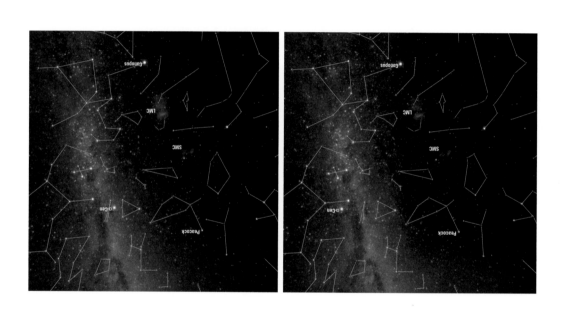

SOUTH CIRCUMPOLAR STARS

4 TO 200,000 LIGHT-YEARS

These stars, always below the horizon from mid-northern latitudes, are seen well from mid-southern latitudes. In Centaurus (the Centaur), Alpha Centauri (α-Cen for short), only 4 light-years away, pops out in the foreground—so much that you have to cross your eyes a bit to focus on it, just as when you focus on your thumb in front of a distant lamp. Just below is the small, famous Southern Cross, made of only four stars. The Milky Way runs through the background at right. The Large and Small Magellanic Clouds (LMC and SMC), respectively 160,000 and 200,000 light-years away, are satellite galaxies of the Milky Way. The six star charts you have just viewed cover the entire sky as visible from Earth. Imagine sitting inside a cubical box looking out in all directions from Earth. The north circumpolar stars form the top of the box. The summer, autumn, winter, and spring stars are the four sides of the box. The south circumpolar stars form the bottom of the box.

GLOBULAR CLUSTER M13

22,000 LIGHT-YEARS

M13 is a *globular cluster*—a spherical, gravitationally bound group of several hundred thousand stars orbiting around the center of our Milky Way. It was discovered by Edmund Halley in 1714. It's called M13 because Charles Messier included it as the 13th object in his famous *Catalogue of Nebulae and Star Clusters* (published in 1771). The reddish stars seen in the image are red giant stars. The blue stars in the central regions are mostly "blue stragglers"—stars recently formed when two stars collide and merge. We are just seeing the cluster's brightest stars here; there are also many red dwarf stars in this cluster that are too faint to see. In 1974, astronomer Frank Drake used the 1,000-foot-diameter Arecibo radio telescope to send a radio message to M13, coded to announce our presence to any extraterrestrials that might live there. It's due to arrive 22,000 years after 1974.

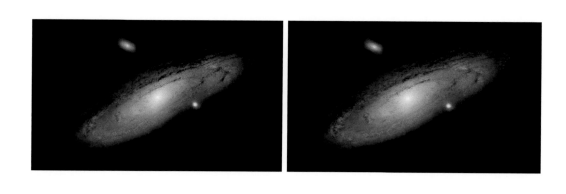

ANDROMEDA GALAXY M31

2.5 MILLION LIGHT-YEARS

The Andromeda galaxy is a companion galaxy to our own Milky Way. (Both are members of our *Local Group of galaxies*.) The astronomer Edwin Hubble found very faint Cepheid variable stars in this nebula. Using them as *standard candles* (per Henrietta Leavitt's calibration of the intrinsic luminosity of such stars), Hubble determined that they were far outside our own galaxy and concluded that this nebula was a large galaxy like our own. The diameter of the Andromeda galaxy is 120,000 light-years (compared to the Milky Way's girth of 100,000 light-years). Like the Milky Way, the Andromeda galaxy (M31) is a *spiral galaxy*—a rotating disk of stars and gas—seen here nearly edge-on, with dust lanes and spiral density waves (moving gravitational traffic jams) of stars circling its center. Two satellite *elliptical galaxies* orbit it: M32 (above and to the left) and M110 (below and to the right). These three galaxies can easily be seen using small telescopes.

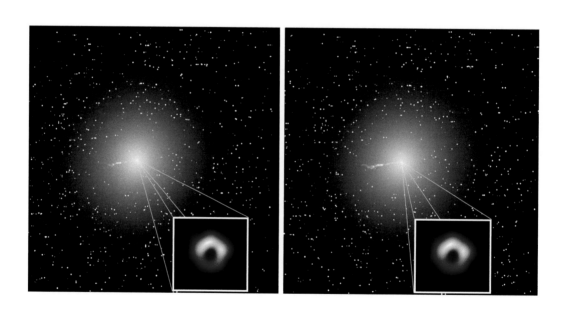

GALAXY M87 AND ITS BLACK HOLE

54 MILLION LIGHT-YEARS

M87, accompanied by its swarm of globular clusters, is a giant elliptical galaxy in the *Virgo cluster of galaxies*. Its light is due to old stars. In the galactic nucleus (at its center) lurks a supermassive black hole, with a mass 6.5 billion times that of the Sun and a radius of 12 billion miles. In 2019, the dark silhouette of the black hole in M87 was photographed against a bright background of infalling gas forming a ring-like accretion disk (see the greatly magnified inset). Gas from the disk's inner edge is being swallowed by the black hole, never to be seen again. Simultaneously, energetic particles are ejected out at the disk's pole forming the 5,000-light-year-long blue jet you see pointed outward toward us. Black holes are common throughout the universe. In 2015, astronomers detected gravitational waves from the inspiral collision of a 29-solar-mass black hole and a 36-solar-mass black hole.

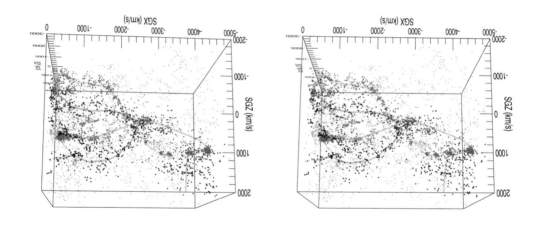

OUR LOCAL UNIVERSE

<300 MILLION LIGHT-YEARS

In this image, you will see the distribution of galaxies (shown as dots) in the local universe. The box dimensions are 243 million light-years (left to right), 194 million light-years (top to bottom), and 291 million light-years (front to back). Clusters of galaxies (red) are connected by filaments made of galaxies traced by colored lines. The Centaurus cluster (red) is in the center, where a number of filaments of galaxies (of different colors) meet. The nearby Virgo cluster is the red cluster at the end of the green filament where it intersects the right wall. The Milky Way galaxy is part of the Virgo supercluster. It is just a little dot at the middle of the right wall of this box. This system of filaments of galaxies connecting clusters of galaxies is known as the *cosmic web*.

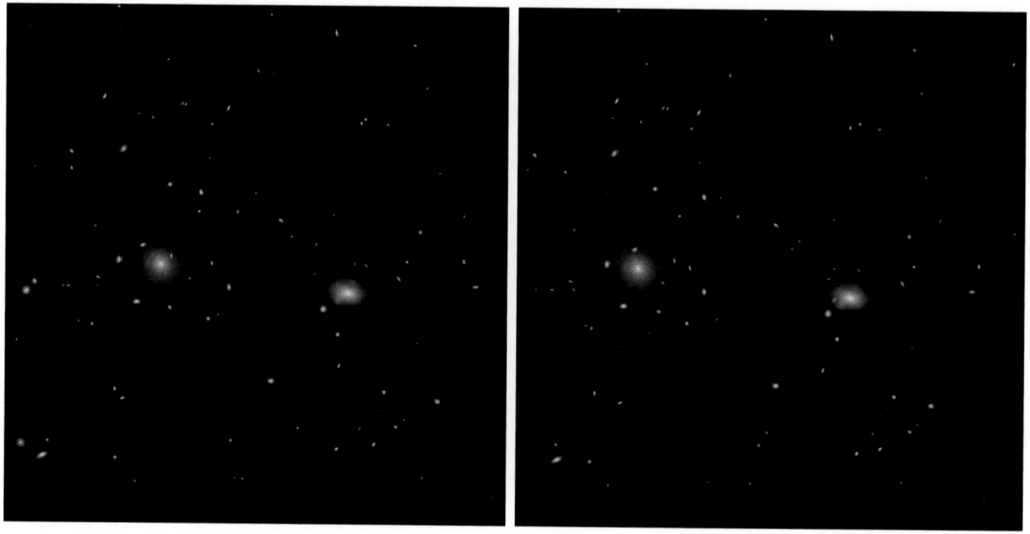

COMA CLUSTER

340 MILLION LIGHT-YEARS

The Coma cluster of galaxies contains over 1,000 galaxies. Here, you get a close-up view of its central portion. In 1934, Fritz Zwicky realized that the high orbital velocities of the galaxies about the center of the cluster implied a total mass much larger than the total mass of the individual, visible galaxies. He suggested the cluster was held together by something invisible—nonluminous *dark matter*. We now believe, from studies of the cosmic microwave background radiation (the ancient remnant radiation left over from the Big Bang) that this dark matter cannot be made of ordinary matter (as found in atoms), but must be in the form of feebly interacting elementary particles (yet to be discovered). Most galaxies in the cluster are elliptical or spiral galaxies stripped of their own gas as they plowed through hot intracluster gas. Prominent here are two giant elliptical galaxies that have gained mass by gobbling up smaller galaxies with which they have collided.

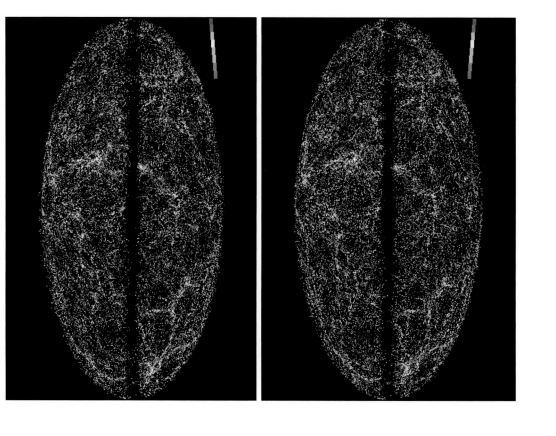

THE COSMIC WEB

<1.0 BILLION LIGHT-YEARS

This all-sky view covers 360° from top to bottom and shows 1.5 million nearby galaxies as dots. The central black band is due to dust along the Milky Way's galactic equator, which blocks our view of distant galaxies. Edwin Hubble discovered that the universe is expanding, such that the spectra of these galaxies are redshifted (Doppler-shifted) according to their distance; the more distant a galaxy is, the faster it is moving away from us. The galaxies in this picture are color coded: red galaxies (the most redshifted) are moving away from us with velocities of up to 7.2% the speed of light, while violet galaxies (the least redshifted) are closest to us. The Virgo cluster of galaxies is a clump of violet dots on the far right in the very foreground. The clusters are connected by filaments in a beautiful glittering structure we call the *cosmic web*.

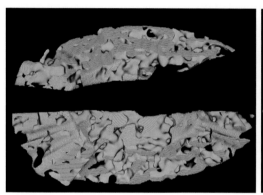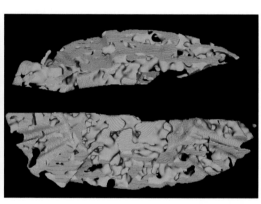

COSMIC WEB'S SPONGE-LIKE NATURE

0.8 TO 1.5 BILLION LIGHT-YEARS

This image, from the Sloan Digital Sky Survey (SDSS), shows the sponge-like nature of the large-scale structure of the universe. We see two lung-shaped survey regions containing 400,000 galaxies. Earth (not shown) would be in the foreground. The solid regions are high density (higher than the median) in terms of the number of galaxies present. These high-density regions form a sponge-like structure, with clusters of galaxies connected by filaments of galaxies—a configuration now called the *cosmic web*. The complementary, interlaced empty regions represent low-density voids connected by low-density tunnels. In 1986, J. Richard Gott, Adrian Melott, and Mark Dickinson proposed such a sponge-like distribution based on the theory of inflation where cosmic-scale structure arises from random quantum fluctuations in the early universe, which grow in magnitude through the action of gravity over 13.8 billion years of cosmic history. Subsequent astronomical surveys such as the SDSS have confirmed the validity of this idea.

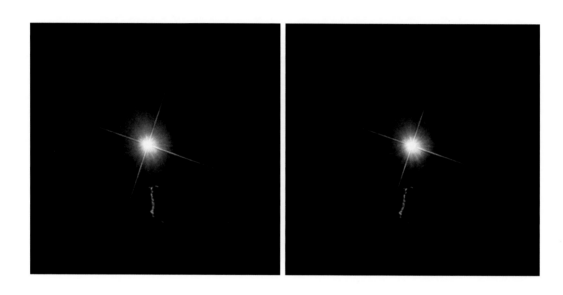

QUASAR 3C273 (AND JET)

2 BILLION LIGHT-YEARS

Quasars are some of the most luminous objects we can see in the universe. Quasar 3C273 is over 4 trillion times the Sun's luminosity, powered by a disk of gas spiraling into a 900-million-solar-mass black hole at the center of a galaxy. It is so bright it simply overwhelms the light from the rest of the galaxy. In 1963, Maarten Schmidt discovered that 3C273's strange spectral features were actually just classic lines of atomic hydrogen, Doppler-shifted to the red, implying it is receding from us at 16% the speed of light. Its recession velocity results from the expansion of the universe. Like M87, it has a jet of particles, shooting out at nearly the speed of light, and extending, in this case, 2 million light-years in our general direction. The quasar image shows four spikes, imaging artifacts from the Hubble Space Telescope's four fin-like secondary mirror supports.

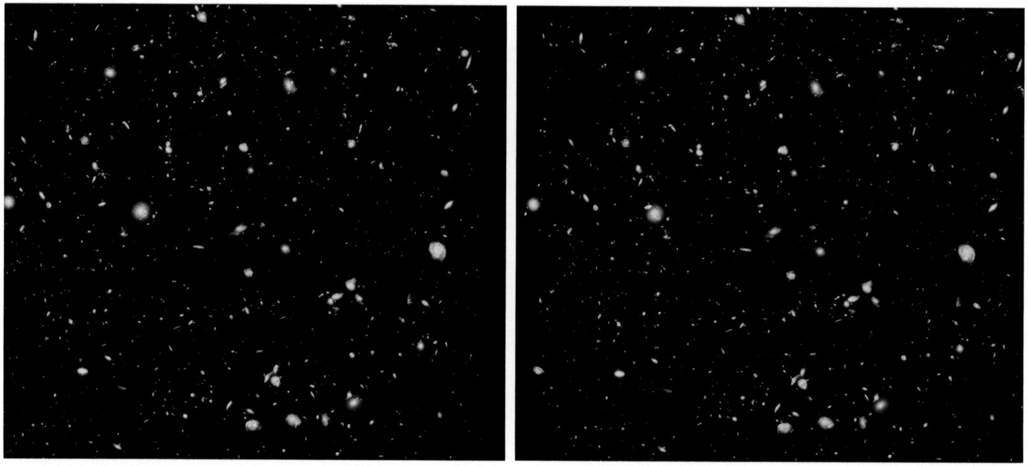

HUBBLE ULTRA-DEEP FIELD

≤13 BILLION LIGHT-YEARS

This striking view of distant galaxies was taken by the Hubble Space Telescope. It covers a tiny region of the sky about 1/25 of a degree across (about the width of a postage stamp 120 feet away). We see only about 1/26 millionth of the entire sky in this picture. Yet, even in that tiny region, there are about 10,000 galaxies. Extrapolating from there, the total number of galaxies in the entire sky within the range of the Hubble Space Telescope is about 260 billion galaxies (10,000 × 26 million). The galaxies that look like specks at the back are about 13 billion light-years away—and, as their light is just reaching us now, we are seeing them as they appeared 13 billion years ago, only about 800 million years after the Big Bang.

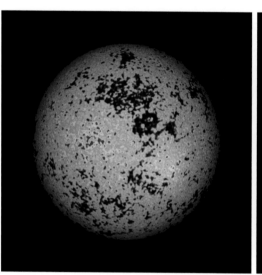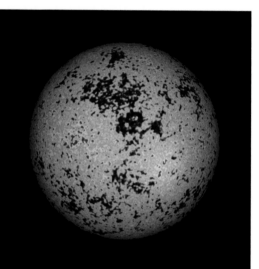

COSMIC MICROWAVE BACKGROUND

13.8 BILLION LIGHT-YEARS

Looking out to a radius of 13.8 billion light-years in all directions, we see the cosmic microwave background (CMB, for short)—light that was emitted soon after the Big Bang beginning of the universe 13.8 billion years ago. It's the most distant thing we can see. Earth is at the center of this spherical shell—illustrated here—and our visible universe lies within. Penzias and Wilson's 1965 discovery of this radiation proved that the universe began with a hot Big Bang. The temperature of the CMB radiation is remarkably uniform (2.73° Kelvin above absolute zero). It has cooled as the universe expanded. Its small fluctuations in temperature (around one part in 100,000) are shown here ranging from red (slightly hotter) to blue (slightly cooler). These fluctuations are in agreement with the theory of inflation, which says the Big Bang began with a hyper-fast period of expansion, in which the universe doubled in size many times during its first 10^{-35} seconds.

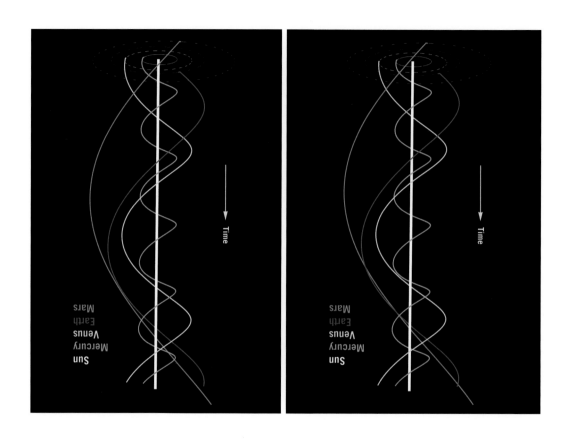

INNER SOLAR SYSTEM SPACETIME

<25 LIGHT-MINUTES

Einstein showed that our universe is four-dimensional, with three dimensions of space (width, depth, height) plus one dimension of time. This 3D *spacetime diagram* shows two dimensions of space (width, depth) horizontally and the dimension of time vertically (arrow pointing to the future). Imagine a movie of planets circling the Sun. Cut the film into frames and stack them on top of each other. Each frame represents an instant in time. The first frame, at bottom, shows a 2D snapshot of the orbits in space as faint-dashed ellipses. The Sun (white) is stationary at the center and so in the stack of frames becomes a vertical white column. That's the Sun's *worldline*. *Earth's worldline*, its path through spacetime, is a *blue helix* of radius 8.3 light-minutes, shown winding once around the Sun in a year in the vertical (time) direction. Mercury and Venus are more tightly wound helixes. Mars, orbiting more slowly, is a loosely wound red helix.

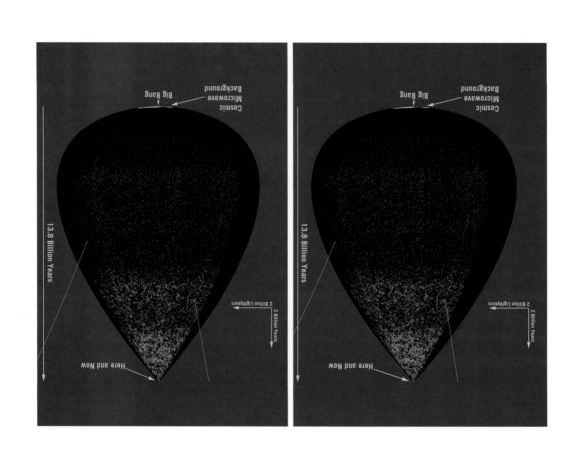

COSMOLOGICAL SPACETIME DIAGRAM

≤13.8 BILLION LIGHT-YEARS

The speed of light is finite. This means the farther out into space we peer, the farther back in time we look, as indicated by the black teardrop-shaped surface in the diagram. Observed galaxies are depicted as green dots; quasars are orange dots. The prominent filament of galaxies near the top is the Sloan Great Wall of Galaxies, measured by J. Richard Gott and Mario Jurič to be a record-breaking 1.37 billion light-years long. Worldlines of a typical galaxy (green line) and quasar (orange line) are shown. Their worldlines bend outward, because the universe is currently expanding more rapidly with time as one goes upward—that is, toward the future. The cosmic microwave background, the most distant thing we can see, is shown near the bottom. The entire visible universe and the worldlines of all galaxies and quasars emerge from the Big Bang—the origin of the universe, 13.8 billion years ago—at the bottom of this diagram.

MAP OF THE UNIVERSE

≤13.8 BILLION LIGHT-YEARS

This picture of the astrocarpet in the front corridor of Peyton Hall, where the Princeton astrophysics department can be found, summarizes our voyage in this book. Walking down the hall, each step takes you 10 times farther away from Earth's center. From left to right, this map offers a 360° panorama of space. Below the line designating Earth's surface, you can see Earth's interior. Above Earth's surface you can see the International Space Station (ISS) and Hubble Space Telescope (HST) in low-Earth orbit. Farther out are the Moon, the Sun, and other planets. More steps down the hall take you to other stars and the prominent wavy gray band of Milky Way stars, snaking from left to right completely across the hall. Beyond are other galaxies. At the very back, in green, is the cosmic microwave background, the most distant thing we can see, marking the edge of the observable universe.

CONCLUSION

We hope this book has given you a new perspective on the universe. We hope it has given you the immersive feeling that you live *in* the universe. Once you have seen the full Moon as it really is—a three-dimensional sphere (rather than a flat disk)—it should look a bit different to you when you see it again in the sky. You will remember what the constellations are really like, the next time you spot them. Perhaps one of our readers, awakened to the sense of being in space, may be inspired to become an astronaut. A new perspective can change our understanding, as when the *Apollo 8* astronauts first glimpsed Earthrise over the lunar horizon—a perceptual shift that arguably helped birth the environmental movement.

This book also provides historical perspective, telling the story of how astronomers discovered the universe in depth—one of the magnificent achievements of human civilization.

First, astronomers discovered the distance to the Moon, by comparing the parallax views of the Moon passing in front of the Sun during a solar eclipse as seen from different places—like having your left and right eyes separated by hundreds of miles.

This measurement, utilizing solar eclipse observations from 190 BC, used the same principle that has allowed you to see the 3D images in this book. It showed the Moon to be 60 Earth radii away—surprisingly far. The first accurate measurements of the distances of the Sun and planets were made in the 1700s by observing, from widely separated locations on Earth, events such as the transit of Venus in front of the Sun. The Sun turns out to be 400 times farther away than the Moon. For measuring the distances to the stars, astronomers took pictures of the sky six months apart, from opposite sides of the Earth's orbit. The nearest star system is Alpha Centauri, 250,000 times as far away as the Sun. The dim stars you see in the sky are suns like our Sun—they are just very far away. This tremendous revelation pushed our idea of the universe far beyond what we had imagined before. The Milky Way, that faint band of light you see from a dark site, is actually the combined light of myriad stars in the disk of our own Milky Way galaxy. The center of our 300-billion-star galaxy is 6,000 times as far away as Alpha Centauri.

Variable stars and exploding stars, serving as "standard candles" to light our way, allowed us to measure their distances by observing how bright they appear. The most distant thing you can see with your naked eye from a dark site is the Andromeda galaxy. It is 100 times farther away than the center of our Milky Way galaxy. The Hubble Space

Telescope lets us see galaxies 5,000 times farther away than the Andromeda galaxy. Over the entire sky, there are 260 billion galaxies within the range of that telescope. The universe is breathtakingly large. We are a tiny dot in the vast cosmos. We can see out to a radius of 13.8 billion light-years in every direction. That distance is called the lookback-time distance. The cosmic microwave background, the radiation left over from the Big Bang, comes directly to us from that radius, a distance over 330,000,000,000,000,000 (330 quadrillion) times as far away as the Moon! Our observable universe fits inside that sphere—all the stars and galaxies we can see. But our universe is far larger than the part we can see. If you go to the top of the Empire State Building, and look out in all directions, you will see a circular region inside the horizon including parts of New Jersey, New York, and Connecticut, a circular region centered on the Empire State Building. If you went to the top of the Eiffel Tower, you would see a circular region including parts of central France centered on the Eiffel Tower. Likewise the observable universe we can see is spherical and centered on the Earth, not because we are at a special place, but because our observations are limited by our viewpoint in space and time—13.8 billion years after the Big Bang. Just as the Earth extends much farther than what can be seen from the top of the Empire State Building, our universe stretches much farther beyond the part we can see. And beyond our universe may lay other bubble universes, floating in a now continually inflating sea of high-energy dark energy, like bubbles in eternally fizzing champagne.

In this vast cosmos, the farthest humans have traveled is to the Moon. This is humbling for sure. It can make you feel small.

But it should also make you feel inspired. For we human beings, though tiny, have figured out how big the observable universe is and our place in it. We live on a planet orbiting an ordinary star in an ordinary galaxy in a typical supercluster of galaxies. We have figured out how old our universe is—13.8 billion years old. Just beyond the most distant galaxies that the Hubble Space Telescope has seen, we have glimpsed the light from the creation of our universe itself—the Big Bang. We may not have gone very far ourselves, but with our space probes and telescopes we have soared outward into the depths of space. We have encountered wonders along the way: a volcano higher than Everest, an ocean more voluminous than Earth's, a storm hundreds of years old, magnificent rings around Saturn, icy denizens of the outer solar system, exoplanets circling other stars, one star 900 times the Sun's diameter and another only 18 miles across, expanding debris from a star that exploded long ago, as well as the elusive black hole. There are giant clusters of galaxies held together by invisible dark matter, and the cosmic web, a sponge-like network of filaments of galaxies connecting great clusters of galaxies.

Many of these wonders have been discovered within one human lifetime. Seventy years ago, we had not yet discovered quasars, pulsars, black holes, exoplanets, the cosmic microwave background, or dark energy, which fills intergalactic space and constitutes 70% of the mass-energy of the universe. The next 70 years promise to be equally interesting. We hope that in addition to giving you an in-depth view of the universe, we have imparted to you a more in-depth understanding of the exciting things still being discovered in our amazing universe.

The vast universe stretches before us. A question facing humanity now is whether, after seeing the view from the Moon, we will as a species turn our backs on the universe and remain on Earth until an extinction event takes us out, or will head out to Mars and beyond. The universe beckons.

We dedicate this book to our descendants Miranda and Travis; Alexandro and Camilla; Elizabeth, Allison, and Jennifer; and Marcy and Diana; as well to our readers who will be exploring that as-yet unexplored region of the universe—the future.

ACKNOWLEDGMENTS

First and foremost, we thank our families for their love and support during the time of writing this book. We thank our wonderful editor Ingrid Gnerlich for her faith in this project; she was also our editor for *Welcome to the Universe: An Astrophysical Tour*. We thank Mark Bellis, our production editor, and Kathleen Kageff, our copyeditor. We also thank Rich's wife, Lucy, for providing her professional editorial help as she did with our previous book.

This book drew its inspiration from many sources. In particular, the book *Beneath the Sea in 3-D* by Mark Blum (Chronicle Books, 1997) does for the ocean depths what we are attempting to do here for the depths of space. We have found personal inspiration from Brian May—from his books *A Village Lost and Found* (Frances Lincoln and the London Stereoscopic Company, 2009) and *Mission Moon 3-D* (with David J. Eicher, London Stereoscopic Company, 2018)—and from his encouraging emails.

We are grateful to our colleagues, friends, and others who have kindly given us permission to use pictures of theirs in this book, including Ricardo Barros, Igor Chekalin, Helène and Denis Courtois, Aram Friedman, Christopher Go, Damian Peach,

Andrew Hamilton, Xavier Haubois, Yehuda Hoffman, Philippe Lamy, David Lane, Daniel Pomarède, and Brent Tully.

Many of the pictures in this book are drawn from the archives of NASA and the European Space Agency (ESA), both of which have the generous policy that all their images are in the public domain. We similarly thank the SpaceX Company and the Sloan Digital Sky Survey for the use of their images. We also thank Chuck Allen for helpful comments.

GLOSSARY

ANGULAR DIAMETER The angle an object spans as viewed by your eye. The Sun and Moon both have angular diameters of about half a degree as seen from Earth. Angular diameters are measured in degrees; there are 360 degrees in a full circle. Degrees are divided into 60 arcminutes and 3,600 arcseconds.

ASTEROID Rocky bodies in the inner solar system most often encountered between the orbits of Mars and Jupiter. Asteroids that cross Earth's orbit would pose a danger if they were to hit us. An asteroid about six miles across that hit the Yucatán Peninsula 65 million years ago caused the extinction of the dinosaurs. A large asteroid such as Ceres, which gravity has forced into a spherical shape, may be designated as a dwarf planet. See also *Kuiper belt object*.

BIG BANG The expanding state at the beginning of our universe, 13.8 billion years ago.

BLACK HOLE A massive object that has become so compressed that nothing, not even light, can escape it. Fall into a black hole, and you can never come back out. Cores of massive stars can collapse into black holes as the stars die. Black holes with masses millions or billions of times the mass of the Sun are found in the centers of most large galaxies.

BLUE STRAGGLER A type of luminous blue star, often found in globular clusters, which is thought to be formed from the collision of two other stars.

BLUE SUPERGIANT A hot massive star in the late stages of its evolution. The star Rigel in Orion is an example of a blue supergiant.

BUBBLE UNIVERSES Low-density expanding bubbles in a high-density inflating sea of dark energy (our universe may have started as one). The entire ensemble is known as a *multiverse*.

CEPHEID VARIABLE STARS Stars that vary in brightness—brighter, dimmer, then brighter again, with characteristic periods that can be used to calibrate them as "standard candles." Henrietta Leavitt discovered that a Cepheid's average luminosity was a function of its period. So if one measures the period of a Cepheid, one can deduce its luminosity. Thus, these stars are "standard candles," like standard streetlights whose luminosity is known, and by observing how bright they appear, one can determine their distance. Cepheid variables are very luminous stars, many over 10,000 times as luminous as the Sun. When Hubble discovered very faint Cepheid variables in the Andromeda nebula, he could use Leavitt's work to determine that they lay far outside our own galaxy, and that the Andromeda nebula was actually another whole galaxy, rather like our own Milky Way.

CHROMOSPHERE A hot region immediately beyond the photosphere of the Sun. The chromosphere of the Sun becomes visible to the naked eye during a total solar eclipse. At no other time should one ever look directly at the Sun.

CIRCUMPOLAR STARS From mid-northern latitudes, Polaris (the Pole star) and the stars that are in its area of sky; always above the horizon as Earth turns.

COMET An icy body entering the inner solar system, which grows a tail due to out-gassing of water vapor and release of dust as the Sun heats it. Most comets originate in the Oort cloud surrounding the solar system; the Oort cloud extends, in distance from the Sun, roughly from 2,000 times to 100,000 times the distance from the Sun to the Earth.

CORONA The hot diffuse halo of gas in the outer atmosphere of the Sun. It has a temperature of roughly 2 million degrees Fahrenheit.

COSMIC MICROWAVE BACKGROUND Light left over from the hot Big Bang that has redshifted into the microwave region of the spectrum.

COSMIC RAYS High-energy particles (including electrons and protons), which are observed to impinge on the Earth's upper atmosphere.

COSMIC WEB The large-scale architecture of the universe: a sponge-like network of filaments (chains) of galaxies connecting great clusters of galaxies, in a giant web.

COSMOLOGICAL CONSTANT A term Einstein added to his field equations in a failed attempt to make a static universe. He abandoned it when Hubble showed the universe is expanding. But Lemaître later explained that the cosmological constant could be understood as what we now call *dark energy*, which is thought to drive the observed accelerated expansion of the universe.

CRATERS Depressions in a body's surface, often due to impacts from smaller asteroids and meteoroids over eons. The many craters seen on the Moon, Mercury, and asteroids are such impact craters. Few impact craters are visible on Earth, as they are eroded away by weather and geological activity.

DARK ENERGY Energy accompanied by negative pressure filling all space. Dark energy is gravitationally repulsive and is thought to be the reason that the expansion of the universe is accelerating.

DARK MATTER Nonluminous matter constituting about 25% of the mass-energy in the universe, which gravitationally holds great clusters of galaxies together. Most of the dark matter in the universe is thought to be composed of as-yet undiscovered elementary particles.

DOPPLER SHIFT Changes in wavelengths of light (or sound) due to an object's moving either toward or away from us. An object moving away from us will have its light shifted to longer wavelengths, that is, redshifted. The wavelengths of light reaching us will be stretched out because each successive wave crest is being emitted from farther and farther away as the object recedes from us.

DWARF PLANET See *planet*.

ECLIPTIC The path on the sky through which the Sun appears to move through the year.

ENERGY DENSITY The amount of energy present per cubic centimeter. Energy is measured in ergs. Alex Filippenko notes an erg is the energy a fly would expend doing one "push up." Dark energy has an energy density of 5.4×10^{-9} ergs per cubic centimeter, which, using $E = mc^2$, is equivalent to 6×10^{-30} grams per cubic centimeter.

EXOPLANET A planet orbiting a star other than the Sun.

GALAXY A conglomeration of stars, gas, and dark matter held together by their mutual gravity. The Milky Way galaxy, in which we live, contains several hundred billion stars and is roughly 100,000 light-years across. There are hundreds of billions of galaxies in the observable universe.

GALAXY CLUSTER A grouping of hundreds to thousands of galaxies.

GENERAL THEORY OF RELATIVITY Albert Einstein's theory explaining gravity in terms of curved spacetime. He developed this theory in 1915, and it has been amazingly successful, explaining the orbit of Mercury, the slight bending of light from background stars measured during total solar eclipses, the phenomenon of gravitational lensing, the nature of black holes, the physics of the hot Big Bang, the expansion of the universe, and the recent detection of gravitational waves.

GLOBULAR CLUSTER A roughly spherical cluster of several hundred thousand old stars, orbiting a galaxy. There are over 100 globular clusters associated with our Milky Way galaxy, while more massive galaxies such as Messier 87 can have thousands of globular clusters.

GRAVITATIONAL LENSING A phenomenon predicted by Einstein's General Theory of Relativity, whereby light passing a massive object (such as a galaxy or a black hole) is bent by the gravity of that object. Galaxies seen behind galaxy clusters are often gravitationally lensed, showing highly distorted and elongated shapes.

GREENHOUSE EFFECT The trapping of heat by the atmosphere of a planet, making its surface hotter than it would be in the absence of an atmosphere.

HABITABLE ZONE The region around a star where the surface temperature of a planet is right for liquid water to exist. Liquid water is thought to be a requirement for the planet to host life.

HIGGS FIELD A field permeating space, having a value at every point in space, that endows particles with mass. The particle associated with the Higgs field, that is, the Higgs particle, was discovered in 2012. You may be more familiar with magnetic and electric fields that permeate space. A compass at a given location reacts to the value and direction of the magnetic field at that location. Light is electromagnetic radiation. The photon is the particle (of light) associated with the electromagnetic field. The Higgs field and Higgs particle are analogous.

HUBBLE-LEMAÎTRE LAW Law stating that distant galaxies have redshifts that are proportional to their distance, showing that the universe is expanding.

INFLATION A hyperfast accelerated expansion in the first 10^{-35} seconds of the early universe that powered the Big Bang.

KUIPER BELT OBJECT Icy body in the outer solar system. Large ones, such as Pluto and Eris, that gravity has forced into a nearly spherical shape, may be designated dwarf planets.

LIBRATION The wobble in the aspect of the Moon's near side that it presents to us as it orbits Earth. This variation in viewpoint allows one to make a 3D stereo picture of the Moon, first done in 1864.

LIGHT-YEAR The distance that light (traveling at a speed of 186,000 miles per second) travels in a year: 5.9 trillion miles. One can also refer to a light-second (186,000 miles), a light-minute (11 million miles), or a light-hour (670 million miles).

METEOROID A small rocky body moving through the solar system. Small ones that enter Earth's atmosphere and burn up, to be seen as shooting stars are called meteors, but if they are large enough, a fragment, called a meteorite, may survive to hit the surface.

MARE A dark region on the Moon's surface, due to massive lava flows triggered by an asteroid impact.

MANTLE A layer in Earth's interior, above its core and beneath its crust, consisting of silicates. Ceres has an icy mantle.

LUNAR ECLIPSE When Earth's shadow falls on the Moon. The Moon makes a dramatic transition over a period of several hours from being completely full to becoming much darker. While it is completely in the shadow, it is illuminated by sunlight refracted through Earth's atmosphere and takes on the ruddy color of Earth's sunsets.

LUMINOSITY The total amount of energy per unit time an astronomical object emits. This is an intrinsic quantity, unlike the apparent brightness, which is smaller the farther away the object is from us.

MILKY WAY See *galaxy*.

MULTIVERSE See *bubble universes*.

NAKED-EYE Referring to an astronomical object that is bright enough to be seen without the aid of binoculars or a telescope. Over the entire sky, there are about 6,000 stars that can be seen with the naked eye under good conditions: no clouds, no moon, and well away from artificial lights.

NEBULA From the Latin word for "cloud," an astronomical object (other than a comet) that appears extended, as opposed to point-like. Many objects we now call "galaxies" were originally termed nebulas before their physical nature was known. Some nebulas, such as the Orion nebula, are gas clouds where stars are being born. Others (such as the Ring nebula and the Crab nebula) are expelled gas from dying stars.

NEUTRON STAR The compact remnant of some types of supernovas. Such a star is composed almost entirely of neutrons and has a mass larger than that of the Sun but

PARSEC The distance such that the parallax angle observed for a star would be one second of arc. It is 3.26 light-years. During the course of the year, as Earth orbits the Sun, a star at a distance of 1 parsec would shift back and forth ±1 second of arc in the sky (1/3,600 of a degree).

PARALLAX The method of determining distance by measuring shifts of foreground objects relative to background objects as seen from different viewpoints; the basis of human stereoscopic vision that gives depth to a scene.

OBSERVABLE UNIVERSE That part of our universe for which there has been enough time since the Big Bang for light to reach us. Our universe is thought to extend much farther than the limits that we can see. See also *bubble universes*.

as *pulsars*.

house. We see these pulsing as the beams pass by us—thus we refer to such objects out beams of light or radio waves along their magnetic poles like a rotating light-is only about 18 miles in diameter. Many neutron stars are rapidly rotating and send

PHOTOSPHERE The apparent surface of a star, from which its light is emitted.

PLANET According to the International Astronomical Union definition, a nonstellar body orbiting a star, a body that is (a) large enough for gravity to have forced it into a spherical or somewhat flattened shape and (b) able to sweep its orbit clear of debris so as to dominate the mass in its orbital area. Eight bodies in the solar system qualify: Mercury, Venus, Earth, Mars, Jupiter, Saturn, Uranus, and Neptune. Smaller objects that satisfy criterion (a) but do not satisfy criterion (b) are termed dwarf planets; examples in our solar system include the asteroid Ceres and some of the larger Kuiper belt objects: Pluto, Eris, Makemake, and Haumea. Moons that orbit planets do not qualify as planets. Bodies more massive than 13 Jupiter masses, and burning heavy hydrogen in their cores, do not qualify and are instead classified as brown dwarf stars. Objects burning ordinary hydrogen and heavier elements are classified as stars, as are dead stars that formerly burned nuclear fuel. Black holes, which may form in various ways, are considered to constitute a separate category.

PULSAR See *neutron star*.

QUANTUM MECHANICS The laws governing the behavior of the very small, whereby an object can simultaneously exhibit wave-like and particle-like properties. Quantum mechanics is important for understanding the properties of electrons in white dwarf stars and neutrons in the interiors of neutron stars, as well as random fluctuations ultimately due to quantum uncertainty effects, which are ultimately responsible for the fluctuations seen in the cosmic microwave background.

QUASAR Luminous disk of accreting matter spiraling into a supermassive black hole at the center of a galaxy.

RADIATION In this book, a general term referring to any sort of light. Visible light is one form of electromagnetic radiation, as are gamma rays, X-rays, ultraviolet light, infrared light, microwaves, and radio waves. (Radiation can also refer to energetic particles emitted by radioactive elements—a completely different phenomenon.)

RED DWARF STAR A star, burning hydrogen in its core, with a mass less than half the mass of the Sun. Such stars are relatively cool and appear red in a telescope. They are small (dwarfs) in contrast to red giants and red supergiants.

RED GIANT A star near the end of its life that has greatly expanded in size after it has exhausted the hydrogen fuel in its core. The Sun will eventually become a red giant, expanding to the size of Earth's orbit.

RED SUPERGIANT After its red giant phase, a massive star in its death throes that has become even bigger. Betelgeuse (with a diameter 900 times larger than the Sun) is an example.

REDSHIFT If an object is moving away from us, the shift of its spectral lines toward the red end of its spectrum. If the object is approaching us, its spectral lines will be shifted toward the blue end of the spectrum (a blueshift). Distant galaxies show redshifts because the universe is expanding. See *Doppler shift*.

SOLAR ECLIPSE (TOTAL) Occurs when the Moon passes in front of the Sun, completely blocking out its bright surface. During a total eclipse, one can see the Sun's corona or outer atmosphere with the naked eye. At all other times, one should never look directly at the Sun.

SPECTRAL LINES Specific wavelengths (colors) of light due to either absorption or emission by particular elements, which can be identified in a spectrum of an astronomical object, useful for discovering the object's composition and its velocity along our line of sight.

SPECTRUM The rainbow band of colors that results when a prism breaks up light from astronomical objects into colors. Different colors represent different wavelengths of light. The entire spectrum consists of gamma rays, X-rays, ultraviolet light, visible light, infrared light, microwaves, and radio waves. See also *radiation*.

STANDARD CANDLE A class of astronomical objects whose luminosities have been calibrated, such that observations of their brightness allow one to determine their distance. Cepheid variable stars are an important example of a standard candle.

STEREOSCOPE A device (such as that built into this book!) for observing a pair of specially made pictures as a single three-dimensional image.

SUNSPOT A region on the surface of the Sun where magnetic field lines emerge. Such regions are somewhat cooler than the rest of the surface of the Sun, thus they appear somewhat darker.

SUPERNOVA An exploding massive star. Supernovas are very rare, perhaps one per century in a large galaxy such as the Milky Way.

TIDAL FORCE The difference in gravitational force between the front and back of an object, which can stretch the object out. Tidal forces from the Moon are responsible for our ocean tides, and tidal forces on Jupiter's moon Io drives extensive volcanism. Tidal forces from Earth have locked the Moon such that it always shows the same face to us.

WHITE DWARF A dead star that has used up its nuclear fuel, expelled its outer layers, and shrunk down to about Earth's size. White dwarfs are often found at the centers of planetary nebulas.

WORLDLINE The path of an object through the three dimensions of space and one dimension of time.

SUGGESTED READING

Brown, M. *How I Killed Pluto and Why It Had It Coming*. New York: Spiegel and Grau/Random House, 2010.

Dickinson, T., and A. Dyer. *The Backyard Astronomer's Guide*. Rev. ed. Buffalo, NY: Firefly Books, 2008.

Eicher, D. J., and B. May. *Mission Moon 3-D*. London: London Stereoscopic Company, 2018.

Ferris, T. *The Whole Shebang*. New York: Simon and Schuster, 1997.

Goldberg, D., and J. Blomquist. *A User's Guide to the Universe*. Hoboken, NJ: Wiley, 2010.

Gott, J. R. *The Cosmic Web*. Princeton, NJ: Princeton University Press, 2016.

Gott, J. R., and R. J. Vanderbei. *Sizing Up the Universe*. Washington, DC: National Geographic, 2010.

Greene, B. *The Elegant Universe*. New York: Vintage Books, 1999.

Hawking, S. W. *A Brief History of Time*. New York: Bantam Books, 1988.

Kaku, M. *Hyperspace*. New York: Doubleday, 1994.

Kirshner, R. P. *The Extravagant Universe*. Princeton, NJ: Princeton University Press, 2002.

Lemonick, M. D. *The Light at the Edge of the Universe*. New York: Villard Books/Random House, 1993.

———. *Mirror Earth*. New York: Walker, 2012.

Levin, J. *Black Hole Blues and Other Songs from Outer Space*. New York: Knopf, 2016.

Ostriker, J. P., and S. Mitton. *Heart of Darkness*. Princeton, NJ: Princeton University Press, 2013.

Peebles, P.J.E., L. A. Page Jr., and R. B. Partridge. *Finding the Big Bang*. Cambridge: Cambridge University Press, 2009.

Poundstone, W. *The Doomsday Calculation*. New York: Little, Brown Spark, 2019.

Rees, M. *Our Cosmic Habitat*. Princeton, NJ: Princeton University Press, 2001.

———, ed. *Universe*. Rev. ed. New York: DK, 2012.

Sagan, C. *Cosmos*. New York: Random House, 1980.

Taylor, E. F., and J. A. Wheeler. *Spacetime Physics*. San Francisco: W. H. Freeman, 1992.

Thorne, K. S. *Black Holes and Time Warps*. New York: Norton, 1994.

Toome, D. *The New Time Travelers*. New York: Norton, 2007.

Trefil, J., and M. Summers. *Imagined Life*. Washington, DC: Smithsonian Books, 2019.

Tyson, N. deG. *Astrophysics for People in a Hurry*. New York: W. W. Norton, 2017.

———. *Death by Black Hole*. New York: W. W. Norton, 2007.

———. *The Pluto Files*. New York: W. W. Norton, 2009.

———. *Space Chronicles*. New York: W. W. Norton, 2012.

Tyson, N. deG., M. A. Strauss, and J. R. Gott, *Welcome to the Universe: An Astrophysical Tour*. Princeton, NJ: Princeton University Press, 2016.

Vilenkin, A. *Many Worlds in One*. New York: Hill and Wang/Farrar, Straus, and Giroux, 2006.

Zubrin, R. M. *The Case for Mars*. New York: Free Press, 1996.

PHOTO CREDITS

Mars ("Blueberries"): NASA.

Mars: NASA/Hubble Space Telescope.

Mars's Moon Phobos: G. Neukum et al., Mars Express/ DLR/ESA.

Comet Churyumov-Gerasimenko: ESA/*Rosetta*/Philippe Lamy. Stereo image taken from https://rosetta-3dcomet .cnes.fr/ as described in the paper by Lamy et al.: https:// arxiv.org/abs/1903.02324.

Comet Lovejoy 2014/Pleiades: Adapted from David Lane.

Asteroid Vesta/Asteroid Ceres: NASA/*Dawn* spacecraft.

Jupiter/Jupiter's Moon Ganymede: Adapted from NASA, Christopher Go.

Mars and Jupiter from Earth: Adapted from Damian Peach [Mars] and Christopher Go [Jupiter].

Jupiter's Moon Io: Adapted from NASA.

Jupiter's Moon Europa: Adapted from NASA.

Saturn: Gott and Vanderbei, *Sizing Up the Universe*/ Hubble Space Telescope.

Saturn's Moon Mimas: Adapted from NASA.

Saturn's Moon Enceladus: Adapted from NASA.

Saturn's Moon Titan: Adapted from ESA.

Outer Solar System: Gott and Vanderbei, *Sizing Up the Universe*.

Uranus, Neptune with Earth for Comparison: Adapted from NASA.

Pluto and Its Main Moon, Charon: NASA/*New Horizons* spacecraft images rocked ±3.5°.

Barnard's Star: Robert J. Vanderbei.

Betelgeuse: Adapted from Xavier Haubois et al., *Astronomy and Astrophysics* 508, no. 2 (2009): 923, with permission © ESA/Observatoire de Paris.

Orion Nebula: NASA/ESA/STScI/P. Cote and E. Baltz.

Ring Nebula: Adapted from NASA/Hubble Space Telescope.

Crab Nebula: NASA/Hubble Space Telescope picture— depth derived from spectroscopic velocity data— Charlebois et al. *Astronomical Journal* 139 (2010): 2083.

Black Hole: Andrew Hamilton.

Globular Cluster M13: Robert J. Vanderbei.

Andromeda Galaxy M31: Igor Chekalin.

Galaxy M87 and Its Black Hole: Adapted from NASA/Hubble Space Telescope/(Inset) Event Horizon Telescope collaboration et al.

Our Local Universe: Adapted from Courtois et al., *Astronomical Journal* 146 (2013): 69.

Coma Cluster: Adapted from Sloan Digital Sky Survey.

The Cosmic Web: Adapted from image courtesy of 2MASS/UMass/IPAC-Caltech/NASA/NSF, the Two Micron All Sky Survey. Image obtained as part of the Two Micron All Sky Survey (2MASS), a joint project of the University of Massachusetts and the Infrared Processing and Analysis Center/California Institute of Technology, funded by the National Aeronautics and Space Administration and the National Science Foundation.

Cosmic Web's Sponge-Like Nature: Adapted from Gott et al., *Astrophysical Journal* 675 (2008): 16.

Quasar 3C273 (and Jet): Adapted from NASA/Hubble Space Telescope.

Hubble Ultra-Deep Field: Adapted from NASA/Hubble Space Telescope.

Cosmic Microwave Background: Adapted from NASA/WMAP.

Inner Solar System Spacetime: Tyson, Strauss, and Gott, *Welcome to the Universe*.

Cosmological Spacetime Diagram: Gott and Vanderbei, *Sizing Up the Universe*.

Map of the Universe: Ricardo Barros.

INDEX